FRAMEWORK FOR DYNAMIC MODELLING OF URBAN FLOODS AT DIFFERENT TOPOGRAPHICAL RESOLUTIONS

Framework for Dynamic Modelling of Urban Floods at Different Topographical Resolutions

DISSERTATION

Submitted in fulfillment of the requirements of
the Board for Doctorates of Delft University of Technology
and of
the Academic Board of the UNESCO-IHE Institute for Water Education
for the Degree of DOCTOR
to be defended in public
on Friday, March 8, 2013 at 10:00 o'clock
in Delft, The Netherlands

by

Solomon Dagnachew SEYOUM

Master of Science Degree in Water Science and Engineering specialization in
Hydroinformatics, born in Addis Zemen, Ethiopia

This dissertation has been approved by the supervisor:

Prof.dr. D. Solomatine

Composition of Doctoral Committee:

Chairman	Rector Magnificus, Delft University of Technology
Vice-chairman,	Rector UNESCO-IHE
Prof.dr. D. Solomatine	Supervisor, UNESCO-IHE/Delft University of Technology
Dr. Z. Vojinovic	Co-supervisor, UNESCO-IHE
Em.Prof.dr. R.K. Price	UNESCO-IHE/Delft University of Technology
Prof.dr.ir. L.C. Rietveld	Delft University of Technology
Prof.dr. S. Djordjević	University of Exeter, United Kingdom
Prof.dr. Philip O'Kane	University College Cork, Ireland
Prof.dr.ir. F.H.L.R. Clemens	Delft University of Technology, reserve

CRC Press/Balkema is an imprint of the Taylor & Francis Group, an informa business

© 2013, Solomon Dagnachew SEYOUM

Published by:
CRC Press/Balkema
PO Box 11320, 2301 EH Leiden, The Netherlands
e-mail: Pub.NL@taylorandfrancis.com
www.crcpress.com - www.taylorandfrancis.com

ISBN: 978-1-138-00048-3 (Taylor & Francis Group)

Acknowledgments

This study has been carried out within the framework of the European research project SWITCH (Sustainable Urban Water Management Improves Tomorrow's City's Health). SWITCH is supported by the European Commission under the 6th Framework Programme and contributes to the thematic priority area of "Global Change and Ecosystems" [1.1.6.3] Contract n° 018530. The data for this research was kindly provided by the UK Environment Agency, Dr Sutat Weesakul from Asia Institute of Technology and Dr Zoran Vojinovic of UNESCO-IHE.

This research would not have been possible without the guidance and the help of several individuals who in one way or another contributed and extended their valuable assistance in the preparation and completion of this study.

First I would like to express my sincere gratitude to Professor Roland Price for his exceptional enthusiasm, support, inspiration and guidance throughout this research. His unyielding willingness to listen to the problems I encountered during the research, the discussion I had with him on the content, the way he stimulated my academic curiosity and his care and support have set an example to my personal and professional life. I sincerely thank you for providing the supervision I really needed. I would like to sincerely thank my supervisor Professor Dimitri Solomatine for taking over from Professor Price when called upon, and for providing valuable inputs, continuous encouragement and overall guidance during the research.

I take this opportunity to express my sincere appreciation and gratitude to my co-supervisor Dr Zoran Vojinovic for his enthusiasm, support and for the pleasant and informal way of working together. You involved me in several of your interesting projects at UNESCO-IHE and as a result I gained valuable experience beyond my research activities and developed a good friendship with you. I also thank you dearly for the efforts you exerted to help me obtain the Dutch visa at the time I had difficulty due to unfortunate circumstances. On the same topic, I would also like to express my sincere appreciation to the late Professor Henk Vonhoff and to Mr Jan Luijendijk for their invaluable effort in helping me enter Dutch territory again for this study.

I am grateful to Professor Damir Brdjanovic and Mr Jan Herman Koster for their patience and for allowing me to complete the writing of my thesis during my lecturer position within the department of Environmental Engineering and Water Technology.

Acknowledgments

I would like to thank Professor Arthur Mynett, who was my supervisor during my MSc research at UNESCO-IHE, for his encouragement, inspiration and moral support as well as for his experienced advice on practical matters related to completing a PhD thesis.

Many thanks go to Arlex Sanchez and many of the other PhD fellows within the Hydroinformatics chair group for the experience we shared together and for the social life I enjoyed in your company. During my research I have had the pleasure to work with several MSc students on their thesis research which gave me the opportunity to develop my understanding of several research topics. Many thanks for the time we shared discussing and learning from each other.

Many thanks also to Million Fekade Woldekidan for being my best friend. Starting from our MSc studies up to the completion of our PhD studies we shared many happy and memorable moments and you made my stay in the Netherlands a pleasant one. I would like to extend my gratitude to my many friends in Ethiopia; in particular I wish to thank Mulugeta Azeze and Molla Ejigu for their enduring friendship. I would also like to thank many of Ethiopian PhD and MSc fellows at UNESCO-IHE for the good times we shared on several occasions.

My parents, my mother Askal Moges and my father Dagnachew Seyoum, deserve special gratitude for their unconditional love, unceasing support, understanding and prayers. My special gratitude also goes to my brothers Dave, Mini, Aschu and Tedu for their constant encouragement, love and understanding and for their moral support. Especially I would like to acknowledge Mini's role in taking care of our parents while the rest of us are far away. I would also like to sincerely thank Yemisrach Mallede Gesse for supporting my endeavour during the time we shared together.

I am greatly indebted to Jeltsje Kemerink for her love, dedication, care and persistent confidence in me. The love and care you have given me provided a conducive environment to complete the writing of my thesis. I am so grateful for having you by my side and I am very much looking forward to share more quality time with you.

Finally, I would like to thank all members of the doctoral examination committee for evaluating this thesis.

Solomon D. Seyoum
Delft, The Netherlands

Summary

Floods are among the most frequent and costly natural disasters in terms of human hardship and economic loss. The impacts of flooding are especially devastating in urban areas as these areas are densely populated and contain vital infrastructures. Urban flood risks and their impacts are expected to increase as urban development in flood prone areas continues and as rain intensity increases as a result of climate change while aging drainage infrastructures limit the drainage capacity in existing urban areas. The increased risk and severe consequence of flooding drives the need for the development of cost-effective flood mitigation strategies as part of sound urban flood management plans. Efficient prediction of characteristics of flood propagation in urban areas is paramount in developing flood mitigation measures. Urban flood modelling attempts to quantitatively describe the characteristics and evolution of flood flows that occur when a large amount of water moves along drainage systems and urban flood plains.

Although high resolution topographic data is essential for detailed prediction of flood flows in urban areas, due to computational demand of hydraulic model simulations on high resolution grids, topographic data is often generalised to a more manageable resolution and floodplain models are built at much coarser resolutions. However, the generalisation of topographic data within urban environments leads to significant changes in the topography due to the smoothing or disregarding of dominant features such as buildings, walls and vegetation. As a result, surface flow models with a lower resolution more likely produce inaccurate flood simulation results than high resolution surface flow models. Several methods are devised in order to keep the information that can be obtained from high resolution topographic data in coarse grid models. This research focuses on the development and application of an urban flood modelling tool to address the problem of capturing small-scale urban features in a coarse resolution two dimensional (2D) model with the aim of improving flood forecasts in geometrically complex urban environments.

For this research a 2D surface flow modelling system is developed based on a non-convective acceleration wave equation for 2D surface flows. The modelling system represents the urban topography using the ground elevations at the centers and boundaries of cells on a rectangular Cartesian grid. The water levels are determined at the cell centers and the discharges (velocities) are determined at the cell boundaries. The alternating direction implicit finite difference procedure is used to solve the governing equations. The modelling system is tested on different case studies and the results demonstrate that the model produced results which are in good agreement with the results of a reference model which implements the full shallow water equations (in this case MIKE21). The tests show that, whilst neglecting convective acceleration terms which lead to local inaccuracies mainly for velocity predictions, the model is cable of predicting flood inundation depths successfully.

As part of this research a 1D-2D coupled modelling system is developed by coupling the developed 2D modelling system with the Storm Water Management Model (SWMM5). The ability of the coupled modelling system to simulate the interaction of underground sewer network flow and above ground free surface flow is studied in two case studies. Coupled models were built to simulate (i) the drainage system of the Segunbagicha catchment and associated floodplains of Dhaka, Bangladesh, and (ii) the drainage system and floodplain of a catchment along Sukhumvit Road in the inner part of Bangkok, Thailand. Even though the developed 1D-2D coupled models suffer from a lack of sufficient field data to calibrate the models and validate the results, it can be concluded based on previous studies and field observations that the coupled model has the capability to simulate the complex interaction between the sewer flow and the surcharge-induced inundation.

To improve flood forecasts in geometrically complex urban environments using coarse grid models, the non-linear relationships between volume and water depth and between flow area and water depth should be taken into account. In an effort to incorporate these non-linear relationships the 2D flood modelling system is modified. The continuity and momentum equations which describe the 2D surface flow modelling system are rewritten in such a way that the volume and area are expressed as non-simple functions of water depth instead of a fixed plane area of the coarse grid. The modified equations make use of the volume-depth and flow-area-depth relationships that can be extracted from fine resolution DEM. The modelling system is developed so that these relationships can be extracted for desired coarse grid size from available fine resolution DEM.
The performance of the modified modelling approach for coarse grid models is tested for one of the case studies. Two models, one with the modified approach taking the non-linear volume-depth and flow-area-depth relationships into account and another one with the standard linear approach, are used to simulate shallow inundation originating from an inflow into an area. Comparing the flood simulation results of the two modelling approaches shows that the modified modelling approach enables the use of coarse grid models without significant loss of accuracy and with the advantage of reducing the computational time.

With the emphasis given to flood peak control measures that aim to facilitate infiltration, incorporating surface water infiltration processes into urban flood simulation models will help to evaluate the effect of the measures on urban flooding. An algorithm to simulate infiltration based on the modified Horton method is incorporated in the 2D modelling system. Results of case studies to simulate rainfall-runoff processes show that the algorithm based on the modified Horton method allows for satisfactory simulation of the infiltration process.

This research focused on developing a methodology for improved urban flood simulation at different topographical resolutions, thus contributing to the field of urban flood modelling in the following ways:

1. Development and testing of a new 2D urban flood modelling system which solves the 2D non-convective wave equations for gradually varying free surface flow with novel features such as the use of iterations to zoom-in on an accurate solution at each double sweep, the ability to halve or double the time step depending on the convergence tolerance of the solution in order to have adaptable time steps for efficient computation, and the ability to begin from a dry cell condition.

2. Development and testing of the 1D-2D coupled modelling system for simulation of the interaction between flows in subsurface pressurized drainage networks and free surface flows in open urban areas.

3. Development of an efficient way of simulating urban floods at a desired coarse resolution grid with a 2D urban flood modelling system which uses information derived from available high resolution topographic data to minimize the errors which might be introduced into the simulation process due to generalization of the topographic data.

4. Incorporation of infiltration processes in the 2D surface flow modelling system to enable assessment of different flood peak control measures.

This thesis concludes with recommendations and suggestions for further research to improve the limitations of the developed approach and reflects on some general aspects of urban flood modelling. Main areas for further research include:

- Determination of appropriate values for roughness coefficients which may represent more accurately the change in effective roughness across coarse grids.
- Representation of storage and flow-area within coarse grids in the form of flow direction dependent storage-depth and flow-area-depth relationships.

Contents

Contents

1   Introduction

1.1 Background

Floods are among the most frequent and costly natural disasters in terms of human hardship and economic loss. In recent years flooding events are becoming more discernible in the media in many places in the world due to their devastating extensive economic damage and unprecedented loss of human life. Particular in urban areas the impact of flooding can be severe because the areas affected are often densely populated and contain vital infrastructure. The frequency of urban flooding is expected to increase due to increasing urbanization, ageing drainage infrastructure and recognised climate change due to human industrial and commercial activities. Urbanization causes previously permeable ground to be more impermeable due to developments producing increased sealing of the surface with concrete and other construction materials with as result a dramatic increase of the amount of rainwater running off the surface into drains and sewers. Especially on locations where further expansions of urban areas take place in floodplains the flood risk is considerably increased. A common feature of sewerage and drainage networks in many cities around the world is that they are old and their condition is unknown. The capacity to drain the surface water is limited and potentially decreases as result of malfunctioning of the system. In addition, in a considerable number of flood-prone urban areas wetter winters and heavier summer showers are expected to put further pressure on our urban drainage networks as result of climate change (Alam and Rabbani, 2007; Douglas et al., 2008; Grum et al., 2006; Parliamentary Office of Science and Technology, 2007).

Hence, urban flooding is a complex and difficult to avoid problem for many cities around the world due to the issues discussed above causing the frequency of devastating floods to be higher and thus increases the risk for human lives and property.

To avoid loss of life and minimize damage and disruption from flooding, sound flood management practices have to be developed and implemented. Such flood management includes actions such as formulating general defence plans, tightening land use planning and utilizing up-to-date technology in flood prevention and protection. The attention given to flood management practices have highlighted the need for better information systems that facilitate sound flood management (Dubrovin et al., 2006).

One way of improving flood management practices is to invest in data collection and modelling activities, which improve our understanding of the functioning of particular systems and enable us to select optimal mitigation measures. In this respect urban flood modelling plays a significant role as it is aims at developing an understanding of a specific urban drainage system and predicting its behaviour for extreme events so that effective solutions to structural and operational problems can be developed and evaluated (Vojinovic and van Teeffelen, 2007),. It provides invaluable information concerning flood hazard zoning, mitigation measures, real-time flood warning systems, and control strategies for optimal use of storage capacity in the systems.. The role of modelling within urban flood management complements the acquisition of data to improve our understanding of the performance of a given drainage network and our ability to control it better, taking into account the particular urban terrain and other contributing factors.

In urban flood modelling, good geometric and topographical data at an adequate resolution play significant role because they provide the modeller with a representation of the land surface and topography which can be used to describe the primary features of the flow paths through the urban area. In this respect, a Digital Terrain Model (DTM) represents one of the most essential sources of information that is needed by flood modellers. A DTM refers to a topographical map consisting of terrain elevations, which is used to characterise the terrain (or land) surface and its properties. DTMs offer the possibility to automatically extract catchment characteristics by creating flow direction and sub-catchment boundary maps (Vojinovic and van Teeffelen, 2007). DTMs are invaluable for the analysis of the terrain topography, the identification of overland flow paths, for setting-up 2D hydraulic models, processing model results, delineating flood hazards, producing flood maps, estimating damages, and evaluating various mitigation measures.

Two-dimensional (2D) models are often the preferred choice to simulate overland flow more accurately than the more familiar one-dimensional (1D) models. Though 1D models have a clear advantage over 2D models in that they require less data and less computation time, they ignore the two dimensional details of hydraulic processes occurring at intersections and, more generally, in modelling flows in the often extremely intricate network of streets and open spaces. Coupling 1D and 2D models enables a more accurate representation of the urban surface and the underground 1D urban drainage network and as such a coupled 1D-2D modelling approach is preferred by many engineers when modelling floods in urban areas. In particular, such an approach can be used to describe the dynamics and interaction between surface and sub-surface systems.

It is now widely recognised that model results and their accuracy are sensitive to the resolution of the source data (Chen *et al.*, 2012; Evans, 2010; Néelz and Pender, 2008). As the spacing of elevation samples increases, land features and shapes that may have a significant impact on flood flow routes may have a too small scale to be accurately represented. As a consequence the surface becomes more generalised and complicated flow patterns due to vegetation, buildings or other man-made structures cannot be accurately modeled. Ideally, a detailed prediction of flood flows over complex topographies such as in urban areas requires hydraulic model simulations with high resolution grids. These should be capable of incorporating the effects of individual buildings and other topographic features that play significant roles in routing and storing the surface flow and are likely to have an influence on the movement of water across the flood plain. However, the associated computational cost may make this approach infeasible, especially when there is a need to simulate many flood event realisations and/or where there is limited time or resources available to perform the computer simulations. In such circumstances, topographic data is often generalised to a more manageable resolution and floodplain models are built at much coarser resolutions such that complicated flow patterns due to individual buildings and other topographic features become sub-grid scale processes. Such coarse grid models are informative, although to a lesser extent than fine grid models (Néelz and Pender, 2008).

In order to use coarse grids to model urban areas accurately, as much information about the fine detailed features such as building, walls, fences and vegetation present within each coarse grid cell should be retained. To reduce the number of inaccurate flood model simulations which cannot include certain features in the generalisation process, methods are devised in order to keep as much detailed information that can be obtained from high resolution topographic data as possible. This research focuses on the development and application of an urban flood modelling tool to address the problem of capturing small-scale urban features in a coarse resolution 2D model with the aim of improving flood forecasts in a geometrically complex urban environment.

1.2 Urban Floods

Throughout history humans have settled along major rivers and coastal areas. The water bodies ensured the access to water for human survival as well as offered economic opportunities such as transport for trading, agriculture, fishery and industrial development. As result these settlements developed into important economic and political centers attracting further investments and an inflow of people. The population increase has led to scarcity of living spaces within these centers and consequently further expansion of the urban areas into low-lying areas that are naturally prone to flooding (De Sherbinin *et al.*, 2007; Watson *et al.*, 1998).

Flooding can occur mainly in two forms namely watershed flooding and tidal flooding. Watershed flooding occurs in response to severe runoff-inducing rainfall over a catchment. Tidal flooding develops when high tides exceed either the top of bank elevation of tidal channels or the crest of dikes. The two types of flooding can occur in conjunction and when they do, extents of flooding usually increase due to backwater effects (Few *et al.*, 2004).

Watershed flooding is a localized hazard that can be considered in two main categories. The first category deals with flash floods which are mostly the product of heavy localized precipitation in a short time period over a given location though can also be caused by the sudden release of water as result of the failure of a dam or levee. In some areas flash floods can also be caused by the sudden release of water held by an ice jam. Flash floods occur within a few minutes to hours after the flood causing event. The second category deals with general floods which are primarily caused by precipitation over a longer time period over a river basin. As such general flooding is a longer-term event that may last for several days (Dabberdt *et al.*, 2000).

A more detailed category of flooding is given by Balmforth et al., (2006). According to Balmforth et al., (2006) flooding of land areas may be divided into four broad categories: coastal flooding, river or fluvial flooding, localised or pluvial flooding and groundwater flooding. Estuarial flooding is a combination of coastal and fluvial flooding. Other less important categories are other causes of flooding arising from operational defects of drainage channels. Often flooding at any one location arises from a combination of the different categories, though this may be difficult to identify precisely. The categories of flooding as described by Balmforth et al., (2006) are given in the following paragraphs.

Coastal flooding arises for a variety of reasons and in addition to the flooding itself can also lead to coastal erosion with associated littoral drift and deposition. Combinations of high tides, atmospheric effects and wave action can lead to severe coastal flooding. In such cases flooding of land areas can be extensive with significant flood depths and high velocities occurring.

Fluvial flooding occurs when rivers overtop their banks and flood the flood Palin. During heavy or prolonged periods of rainfall river levels rise and may exceed the bank top so that water spreads out alongside the main river channel over what is known as the flood plain. The floodplain naturally provides two functions, conveyance of the additional flow, and storage that attenuates some of the flood effects. Urban development can severely affect this natural process by removing important areas that previously served to convey and attenuate floods.

Pluvial Flooding is caused by the effects of localised heavy rain generating surface runoff beyond the capacity of the drainage network. It can occur in both rural and urban areas, though its effects are more pronounced and damaging in the latter. Pluvial flooding may be directly due to overland flow from land saturated by heavy rain. This is more common during long periods of rainfall in winter months, though it also occurs in urban areas during intense summer rainfall. Pluvial flooding can also occur where local drainage channel capacity is exceeded. It can also occur where there is adequate drainage channel capacity but flow cannot enter the channel at the necessary rate. A good example of this is highway flooding caused by a lack of gully capacity.

Groundwater flooding arises as a result of high water table levels leading to the formation of springs that directly flood land areas and property. It most frequently occurs in winter months after prolonged periods of rain. It generally occurs in areas which are underlain by permeable soil or rock, typically in Chalk, Sandstone or Limestone. Groundwater flooding can be extensive, of long duration, involve high volumes of flood water and is very difficult to control. High water levels can also cause basement flooding through infiltration whilst also increasing base flows in drainage systems through infiltration. Although important it only accounts for a relatively small proportion of urban flooding incidents.

Other causes of flooding due to operational problems such as channel blockage or collapse. It most frequently occurs in sewers and culverts subject to high levels of sediment or tree route ingress, or due to partial or complete collapse. Another common cause is the blockage of trash screens at the entrance to culverts.

Floods in urban conditions are flashy in nature and occur both on urbanised surfaces (streets, parking lots, yards, parks) and in small urban creeks that deliver water to large water bodies (Andjelkovic, 2001). Urban flooding often occurs where there has been extensive development within stream floodplains. Urbanization increases the magnitude and frequency of floods by increasing the impermeable surfaces, and therefore increasing the speed of drainage collection, reducing the carrying capacity of the land and, occasionally, overwhelming sewer systems. Most urban floods can be categorized as general floods and are often caused by severe thunderstorms or rainstorms proceeded by a long-lasting moderate rainfall that saturates the soil.

Urban flooding causes considerable damage and disruption with serious social and economic impacts. Flood damage as defined by Messner et al. (2007) refers to all varieties of harm caused by flooding and it encompasses a wide range of harmful effects on humans, their health and their belongings, on public infrastructure, cultural heritage, ecological systems, industrial production and the competitive strength of the affected economy. Flood damage can be categorised in direct and indirect damages as well as in tangible and intangible damage.

Direct flood damage covers all varieties of harm which relate to the immediate physical contact of flood water to humans, property and the environment. This includes, for example, damage to buildings, economic assets, loss of standing crops and livestock in agriculture, loss of human life, immediate health impacts, and loss of ecological goods. Direct damage is usually measured as damage to stock values. Indirect flood damage is damage caused by disruption of physical and economic linkages of the society, and the extra costs of emergency and other actions taken to prevent flood damage and other losses. This includes, for example, the loss of production of companies affected by the flooding, induced production losses of their suppliers and customers, the costs of traffic disruption or the costs of emergency services. Indirect damage is often measured as the loss of flow values.

Tangible damage refers to damages which can be easily specified in monetary terms such as damage to assets, loss of production, whereas intangible damage refers to casualties, health effects or damages to ecological goods and to other kind of goods and services which are not traded in a market and therefore more difficult to assess in monetary terms (Messner *et al.*, 2007).

1.3 Urban flood management

Urban flooding has become an increasingly important problem and growing issue around the world (Jha *et al.*, 2011). Since it continues to be regarded as an almost inevitable danger, the development of cost-effective flood mitigation strategies has become of the utmost importance for many urban areas.

Andjelkovic (2001) states that total flood protection is unrealistic and unwise and therefore the ultimate goal of flood loss prevention should be the improvement of the quality of life by reducing the impact of flooding and flood liability on individuals, as

6

well as by reducing private and public losses resulting from the flooding. According to Andjelkovic (2001), the main objectives of urban flood management are to provide answers to the questions how to deal effectively with the possibility of flooding in an urban environment and how to cope with the associated uncertainties?

Ahmad and Simonovic (2006) divide and elaborate the flood management process into the following three phases: (i) pre-flood planning; (ii) flood emergency management; and (iii) post-flood recovery. The authors elaborate that during the pre-flood planning phase structural and non-structural flood management options are analyzed and compared for possible implementation to reduce flood damages. Hydrodynamic modelling and economic analysis tools play an important role in this phase. Future population and projections of economic activities are also important in analyzing the long-term impact of decisions made during this phase of the flood management process.

The flood emergency management phase involves the forecasting of floods on a regular basis. A frequent assessment of the current flood situation and the operation of flood control structures are important during this phase. At this stage, urgent decisions are made to protect communities and capital works. This may involve upgrading flood protection works, such as strengthening and extending dikes. Based on an appraisal of the current situation, inhabitants are evacuated from affected or threatened areas.

The post flood recovery phase involves decisions regarding the return to normal life activity from a period of flooding. Some issues of main concern during this phase of the flood management process include the provision of assistance to flood victims, an evaluation of flood damages and the rehabilitation of damaged properties. During this phase, the flood impacts are evaluated and mitigation strategies are implemented.

To reduce the impact of floods damage reduction measures play an important role within urban flood management. Flood damage reduction measures consists of two basic techniques – structural and non-structural (Jha et al., 2011). Structural measures emphasize the construction of physical structures for the purpose of detaining or retaining floods and releasing the stored water at non-damaging rates. Structures can also be built to prevent inundation from floods by diverting flows during peak events or by decreasing or delaying runoff. Amongst others, structural measures include the construction of reservoirs and/or detention basins upstream of the protected area, construction of dykes, construction of drainage and pumping facilities, and physical watershed improvements.

Non-structural measures emphasize the use of regulations, guidelines and disaster preparedness to minimize flood damage. They rely on action and support from households and local organizations working collaboratively, and require the participation of inhabitants in areas prone to flooding. Inappropriate regional and urban planning can exacerbate the negative effects of extreme hydrological events (Barredo *et al.*, 2005). Therefore sustainable land management and planning practices, including appropriate land use planning and development restrictions in flood-prone areas, are introduced as suitable non-structural solutions to minimise flood damage. Other non-structural measures include flood insurance to recognize the risks of floods and to provide compensation when damages are unavoidable with acceptable premiums, protection of individual properties, flood warning systems to evacuate residents and disaster preparedness in order to prepare the community for an effective response in case of an emergency.

Typically, both types of measures, structural and non-structural, are applied to manage urban flood problems as they often complement each other. Structural measures often provide long-term flood protection, though there are technical and economic constraints on the provision of structural measures to control urban flooding. Structural measures are usually expensive and their implementation requires time. On the other hand non-structural measures often involve complex coordination processes as well as issues related to enforcement of regulations. Therefore it is normally important to adopt and balance both kinds of measures for a specific situation in order to achieve an optimal solution to the problems.

Flood management is one of the important aspects of urban storm water management. Urban storm water management involves the development and implementation of a combination of structural and non-structural measures to reconcile the conveyance and storage functions of a stormwater system as well as the related needs of an expanding urban population with the purpose of water conservation, pollution prevention, and ecological restoration. Sustainable strategies for urban stormwater management can be characterized by the following five goals (Chocat *et al.*, 2001);

    i.   flood reduction to minimise peak discharge rates from urban catchments,

    ii.   pollution minimisation by collecting and managing pollution loads generated in urban catchments,

    iii.   stormwater retention (harvesting) and beneficial use of storm water runoff within or near the contributing catchment,

    iv.   urban landscape improvement for instance by incorporating the water into functional green belts; and,

     v.   reducing drainage investments for example through integration of stormwater into green areas and thereby reducing the cost of infrastructure.

Analysis of flood characteristics such as flood water depth, velocity (speed and direction of flood propagation), discharge, timing and duration of flooding are usually incorporated into different steps within the flood management process such as flood hazard assessment, risk analysis and disaster management. Urban flood models play an important role in providing information on the flood characteristics in order to better understand the reasons behind flooding events and find adaptive solutions to reduce the impact of urban flooding. Moreover, urban flood models serve to develop an understanding and to generate predictions of the behaviour of urban drainage systems so that effective solutions to structural and operational problems can be developed and evaluated.

1.4   Urban flood modelling

There is nowadays a range of modelling approaches available for urban floods. Traditionally, urban drainage systems have been modelled using a 1D approach which is limited to modelling flows within channels and/or pipes. A step forward has been the dual drainage concept, where urban surface is treated as a network of open channels and ponds (major system) connected to the channel/pipe system (minor system), commonly referred to as 1D-1Dapproach. Recently, coupled 1D-2D models have emerged, in which channel/pipe network flow models are tightly coupled with the flood flow model that treats the urban surface as a two-dimensional flow domain. In such an approach, complex interactions that take place through surface/sub-surface linkages are explicitly taken into account using appropriate equations. Therefore, urban flood modelling practice concerns the use of 1D, 1D-1D, 2D, and 1D-2D modelling approaches to represent the greater majority of processes occurring within the drainage systems and urban floodplains (see for example: (Chen *et al.*, 2005; Djordjević *et al.*, 2005; Hsu *et al.*, 2000; Mark *et al.*, 2004; Mignot *et al.*, 2006; Neal *et al.*, 2009; Verwey *et al.*, 2008; Vojinovic and van Teeffelen, 2007; Vojinovic *et al.*, 2006; Vojinovic *et al.*, 2011).

The appropriate level of modelling for the assessment of particular issues depends crucially on the nature of the physical situation and on the availability of data. Where flood flows are confined to well-defined conduits, a robust 1D model can usually be instantiated and, once adequately calibrated and verified, its results may be considered reasonably safe for decision-making. However, the flows generated along urban areas are usually highly complex because the topography of the urban surface is eminently artificial with correspondingly highly irregular geometries, and the flows may run contrary to flow paths in a corresponding natural terrain. Such issues necessitate the coupling of simulations using 1D and 2D modelling systems (Vojinovic *et al.*, 2008).

Simulation using coupled 1D-2D models is a rather complex process, and as such it can take a considerable amount of computational time. These simulations are based on complex numerical solution schemes for the computation of water levels, discharges and velocities. The surface model (i.e., 2D model) simulates vertically-integrated two-dimensional unsteady flow given the relevant boundary and other ancillary conditions (e.g., resistance coefficients.) and bathymetry/terrain as provided by a digital terrain model of the catchment area.

1.5   Research Objective

The overall objective of this research is to develop and test a framework for the dynamic modelling of urban flood at different topographical resolutions. This is achieved through the following steps;

1. Developing and testing a new 2D surface flow modelling system for simulation of urban flooding;

2. Developing and testing an algorithm for coupling the SWMM5 1D modelling system with the 2D surface flow modelling system;

3. Create a methodology that enables an automated generalisation of high resolution topographical data for urban flood simulation based on a coarse grid that represents some important consequences of small scale urban features on flood characteristics at the coarser resolution, and;

4. Developing and testing a methodology to enable an improved simulation of urban floods using the 2D urban flood model at different topographical resolution.

1.6   Research Methodology

This research builds on a comprehensive review of available literature on subjects related to urban flood modelling, model development and application. Amongst others the literature review includes the following topics:

- The governing equations used to describe flow of water during a flood event including gradually varying free surface flow, pressurized flow and a combination of gradually varying free surface and pressurized flow;

- The equations governing gradually varying free surface flow in 3D and 2D and Saint Venant equations which govern 1D gradually varying free surface flow including the basic assumptions for their analytical derivations;
- The construction of various flood models depending on which of the various forces in the momentum equations are considered negligible in comparison with the remaining terms;
- The numerical schemes and numerical grids which can be used to solve the conservation equations for mass and momentum and fundamental criteria used to quantify the performance of the numerical techniques in generating solutions;
- The cause and types of urban flooding and available approaches for modelling urban floods, and
- Topographic data, which is one of the most important data items required to build flood inundation models, its source, the spatial resolution, the generation of the Digital Terrain Model, and the incorporation of fine scale features and their effect on urban flood modelling.

As a first step a new 2D surface flow modelling system is developed in order to test a new approach for improving the application and accuracy of 2D urban flood models in different topographical resolutions. To assess its applicability the modelling system is tested on a range of hypothetical case studies with different topographical features. As a next step the 2D surface flow modelling system is coupled with a 1D sewer network modelling system (SWMM5) in order to be able to simulate the complex nature of the interactions between surcharged sewers and flows associated with urban flooding. For this purpose the source code of SWMM5 is modified such that the surcharge in the sewer network is represented in terms of hydraulic head rather than overflow volumes such that there is a dynamic two-way linkage between the surface flow modelling system and SWMM5. The coupled 1D-2D modelling system is tested on two case studies located in the flood-prone cities Dhaka, Bangladesh, and Bangkok, Thailand.

The methodology to improve the application and accuracy of the 2D surface flow modelling system at different topographical resolutions requires the modification of the equations for the surface flow modelling system by integrating them over a coarse grid cell. Accordingly the surface flow modelling system is modified and the proposed methodology is tested using a case study to show its applicability and to compare the results with the standard approach of coarsening the topographical resolution.

The research presented here has been carried out as part of the EU-funded SWITCH project at UNESCO-IHE Institute for Water Education. Beyond the scope of this research, it is planned to develop the urban flood modelling system as an open source application with a user friendly graphical interface in order to facilitate its use and further development by stakeholders and software developers.

1.7 Dissertation Structure

This research contains eight chapters. Chapter two provides comprehensive literature review of topics relevant to the area of urban flood modelling. It describes the general forms, underlying assumptions, simplifications and applicability of the governing equations used to describe the flow of water with a gradually varying free surface. It reviews different types of modelling systems used for modelling shallow water flows and the numerical schemes used to solve the shallow water equations. It also discusses the issues concerning the use of topographical data for urban flood modelling and the treatment of rainfall intensity and infiltration rates in urban flood modelling.

Chapter three provides detailed description of the formulation and development of the modelling systems used in this research. It describes a newly developed 2D modelling system which employs an Alternating Direction Implicit (ADI) numerical procedure in combination with an iteration procedure, development of a coarse grid 2D modelling system which uses information derived from a fine grid resolution for the purpose of improving flood forecasts in geometrically complex urban environment, the coupling of the developed 2D modelling system with a 1D sewer network modelling system (SWMM5) to simulate the complex nature of the interaction between surcharged sewer and flows associated with urban flooding and finally the incorporation of infiltration process in the 2D urban flood modelling system.

Chapter 4 demonstrated the applicability of the 2D urban flood modelling system to simulate urban drainage problems. Chapter 5 covers the case studies carried out to simulate the complex interaction between a one dimensional sewer network and the two dimensional above ground surface flow in urban areas. Chapter 6 is devoted to demonstrate the use of the modified 2D modelling system for the purpose of improving flood forecasts in geometrically complex urban environment using coarse grid 2D urban flood modelling system. Chapter 7 demonstrates the application of the 2D urban flood modelling system with infiltration process. The thesis concludes in chapter 8 with summary of the overall research, and conclusions and recommendations for application and research on urban flood modelling.

2   Literature Review

2.1   Introduction

This chapter sets out to highlight the general forms, underlying assumptions, simplifications and applicability of the governing equations used to describe the flow of water with a gradually varying free surface. It describes different types of models used for modelling shallow water flows and the numerical schemes used to solve the shallow water equations. It also discusses the issues concerning the modelling of urban flooding and urban flood models, and the use of topographical data for urban flood modelling. Finally the chapter reviews the treatment of rainfall intensity and infiltration rates in urban flood modelling.

2.2   Governing Equations

The hydraulics behind the flow dynamics of flooding are represented by particular equations used in hydraulic models for flood modelling. It is of paramount importance that these equations should be properly understood in order to be able to identify the strength and weakness of the hydraulic model, to control and minimize errors and to critically assess the simulation results from a model. Three types of flow can occur in a flooding event, namely gradually varying free surface flow, pressurized flow and a combination of gradually varying free surface and pressurized flow. The gradually varying free surface flow is a gravity driven flow of water under atmospheric pressure. Pressurized flow is flow under pressure in a confined closed conduit. Each flow type is characterized by its own governing equations depending on the underlying assumptions and level of simplifications to define their applicability.

2.2.1   The gradually varying free surface flow

The Navier-Stokes equations form the basis of a general model which can be used to simulate the flow of water in many applications. However, when considering a specific problem such as shallow-water flows in which the horizontal scale is much larger than the vertical one, the Shallow Water Equations, which can be derived from the Navier-Stokes equations by integrating them over the depth of flow, will suffice (Alcrudo, 2004).

A general fluid-flow problem involves the prediction of the distribution of different quantities: the fluid pressure, the temperature, the density and the flow velocity. With this general objective, six fundamental equations need to be considered: the Continuity

Equation based on the law of conservation of mass, the Momentum Equations along three orthogonal directions (derived from Newton's second law of motion), the Thermal Energy Equation obtained from the first law of thermodynamics, and the equation of state, which is an empirical relation between fluid pressure, temperature and density (Aldrighetti, 2007).

Assuming both density and temperature as constants the last two equations can be ignored leaving the Continuity Equation and by the Momentum Equations as the basis of the model.

The set of governing equations most frequently used in 1D free-surface flow are the Saint Venant equations. These equations were named after a French mathematician who first derived them in 1871. The Saint Venant equations are the set of equations that define gradually-varied unsteady flow with the condition that the pressure distribution is hydrostatic. These equations can be derived by using the system-control-volume transformation (the Reynolds transport theorem) (Vreugdenhil, 1994). A more general set of equations can be derived by considering a non-hydrostatic pressure distribution, known as the Boussinesq equations. These can be obtained by integrating the full three-dimensional Navier-Stokes equations (named after Louis Navier, a French physicist, and George Stokes, a British physicist) in the vertical direction, and by making use of the Boussinesq assumption. In this assumption, the flow velocity in the vertical direction is considered to vary from a minimum of zero at the bottom of the flow domain to a maximum at the free surface (Chaudhry, 1993).

The basic assumptions for the analytical derivation of the Saint Venant Equations are the following:

- The flow is one-dimensional, i.e. the velocity is uniform over the cross-section and the water level across the section is horizontal
- The streamline curvature is small and the vertical accelerations are negligible, so that the pressure can be taken as hydrostatic
- The effects of boundary friction and turbulence can be accounted for through resistance laws analogous to those used for steady state flow
- The average channel bed slope is small so that the cosine of the angle it makes with the horizontal may be replaced by unity.
- The flow is gradually varied.
- The Coriolis and wind forces are neglected.
- The liquid is incompressible and homogeneous.

These hypotheses do not impose any restriction on the shape of the cross-section of the channel and on its variation along the channel axis, although the latter is limited by the condition of small streamline curvature.

2.2.2   The 3D Shallow Water Wave Equations

The governing three dimensional primitive variable equations describing constant density, gradually varying free surface flow of an incompressible fluid are the well known Reynolds-Averaged Navier-Stokes Equations which express the conservation of mass and momentum (Aldrighetti, 2007). Such equations have the following form:

$$\frac{\partial u}{\partial x} + \frac{\partial v}{\partial y} + \frac{\partial w}{\partial z} = 0 \tag{2.1}$$

$$\frac{\partial(\rho u)}{\partial t} + \frac{\partial(\rho u^2)}{\partial x} + \frac{\partial(\rho uv)}{\partial y} + \frac{\partial(\rho uw)}{\partial z} = \frac{\partial(\tau_{xx} - p)}{\partial x} + \frac{\partial \tau_{xy}}{\partial y} + \frac{\partial \tau_{xz}}{\partial z} \tag{2.2}$$

$$\frac{\partial(\rho v)}{\partial t} + \frac{\partial(\rho uv)}{\partial x} + \frac{\partial(\rho v^2)}{\partial y} + \frac{\partial(\rho vw)}{\partial z} = \frac{\partial \tau_{xy}}{\partial x} + \frac{\partial(\tau_{yy} - p)}{\partial y} + \frac{\partial \tau_{yz}}{\partial z} \tag{2.3}$$

$$\frac{\partial(\rho w)}{\partial t} + \frac{\partial(\rho uw)}{\partial x} + \frac{\partial(\rho vw)}{\partial y} + \frac{\partial(\rho w^2)}{\partial z} = \frac{\partial \tau_{xz}}{\partial x} + \frac{\partial \tau_{yz}}{\partial y} + \frac{\partial(\tau_{zz} - p)}{\partial z} \tag{2.4}$$

where $u(x, y, z, t)$, $v(x, y, z, t)$ and $w(x, y, z, t)$ are the velocity components in the horizontal x, y and in the vertical z-directions. t is the time, p is the normalized pressure, that is the pressure divided by the constant density, g is the gravitational acceleration and τ is shear stress.

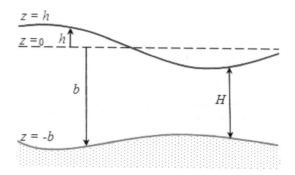

Figure 2-1: Typical water column

In Figure 2-1 $h = h(t, x, y)$ is the elevation (m) of the free surface relative to the datum, $b = b(x, y)$ is the bathymetry (m), measured positive downward from the datum and $H = H(t, x, y)$ is the total depth (m) of the water column. Note that $H = h(t, x, y) + b(x, y)$.

2.2.3   The 2D Shallow Water Wave Equations

From the fully three dimensional equations, it is possible to derive a simplified 2D model assuming that the circulation of interest takes place in the horizontal $x - y$ plane and the three dimensional equations integrated over depth. Detail derivation of the 2D shallow water equations are given in (Vreugdenhil, 1994).

Combining the depth-integrated continuity equation with the depth-integrated x- and y-momentum equations, the 2D (nonlinear) SWE in conservative form are given in Eq. (2.5) to Eq. (2.7):

$$\frac{\partial H}{\partial t} + \frac{\partial}{\partial x}(H\bar{u}) + \frac{\partial}{\partial y}(H\bar{v}) = 0 \tag{2.5}$$

$$\frac{\partial(H\bar{u})}{\partial t} + \frac{\partial}{\partial x}(H\bar{u}^2) + \frac{\partial}{\partial y}(H\overline{uv}) = -gH\frac{\partial h}{\partial x} + \frac{1}{\rho}\left[\tau_{sx} - \tau_{bx} + F_x\right] \tag{2.6}$$

$$\frac{\partial(H\bar{v})}{\partial t} + \frac{\partial}{\partial x}(H\overline{uv}) + \frac{\partial}{\partial y}(H\bar{v}^2) = -gH\frac{\partial h}{\partial y} + \frac{1}{\rho}\left[\tau_{sy} - \tau_{by} + F_y\right] \tag{2.7}$$

2.2.4   The 1D Saint Venant Equations

The one dimensional equations for unsteady flow in open channel can be derived by integrating Equations (2.5) and (2.6) laterally.

Neglecting eddy losses, Coriolis force, atmospheric pressure and wind shear effect, the 2D Saint Venant equations, also known as the Shallow Water equations can be written in conservative form, as a function of the flow area and discharge or the non-conservative form, functions of depth and velocities. In theory, only a strict processing of conservative equations allow one to properly take into account discontinuities, which can appear in the solutions of shallow water equations (such as a hydraulic jump). Other reasons (including the stability of numerical schemes, the use of the method of characteristics) have led to the adoption of a different formulation called the "depth-velocity formulation" (Hervouet, 2007). Depending on the particular problem being considered and the numerical technique being used, it may be more appropriate to deal with one particular form of the equations than another.

In the presence of a water intake or release at the free surface, inside the flow domain or at the bottom, or if one takes into account rain or infiltration to the ground, the right hand side of the continuity equation is not zero but equal to a term that can be a source or sink denoted here as q.

The conservative form of the 2D Saint Venant equations can be written as:

Continuity equation:

$$\frac{\partial h}{\partial t} + \frac{\partial (hu)}{\partial x} + \frac{\partial (hv)}{\partial y} = q \tag{2.8}$$

Momentum equations:

$$\frac{\partial (hu)}{\partial t} + \frac{\partial (hu^2)}{\partial x} + \frac{\partial (huv)}{\partial y} + gh\frac{\partial h}{\partial x} + gh\left(s_{fx} - s_{bx}\right) = 0 \tag{2.9}$$

$$\frac{\partial (hv)}{\partial t} + \frac{\partial (huv)}{\partial x} + \frac{\partial (hv^2)}{\partial y} + gh\frac{\partial h}{\partial y} + gh\left(s_{fy} - s_{by}\right) = 0 \tag{2.10}$$

where u and v are the velocity components on the x and y directions, s_{bx} and s_{by} are the bed slopes on the x and y directions, s_{fx} and s_{fy} are the friction slopes on the x and y directions. The friction slopes are intended to model effects due to boundary friction and turbulence. Their description is rather empirical and developed for use with steady state flow. They are represented as $s_{fx} = c_f u\sqrt{u^2 + v^2}$ and $s_{fy} = c_f v\sqrt{u^2 + v^2}$. The coefficient c_f appearing in the friction terms is normally expressed in terms of the Manning n or Chézy roughness factors

The propagation of flood waves is controlled by the balance of the various forces included in the momentum equations. The first terms in (2.9) and (2.10) represent the local inertia (or acceleration), the second and third terms represent the convective acceleration, the fourth terms represent the pressure forces and the last two terms account for the friction and bed slope respectively. Various flood flow models can then be constructed, depending on which of these terms is assumed negligible in comparison with the remaining terms (Hunter *et al.*, 2007)

Dividing Eq. (2.9) by the gravitational acceleration, g, and the water surface elevation, h, the different types of flood flow model and the terms used to describe them can be written as

$$\frac{1}{g}\frac{\partial u^2}{\partial x}+\frac{1}{g}\frac{\partial(uv)}{\partial y}+\frac{1}{g}\frac{\partial u}{\partial t}+\frac{\partial h}{\partial x}+\underbrace{\left(S_{fx}-S_{bx}\right)}_{\substack{Kinematic}}=0 \tag{2.11}$$

The kinematic, diffusive and non-inertia wave models include the terms as shown in (2.11) and the dynamic wave model include all the terms. These equations can thus be thought of as a progression in complexity for modelling shallow water flows (Hunter *et al.*, 2007)

Dynamic wave models

These models include all the terms of the reduced Saint Venant Equations which are a hyperbolic system of conservations laws. Dynamic models allow for the modelling of a full transient phenomenon. In conservation form, the 2D Saint Venant equations may be more conveniently written in a vector form as;

$$\mathbf{u}_t+\mathbf{F}_x+\mathbf{G}_y=\mathbf{R} \tag{2.12}$$

with the vectors defined by

$$\mathbf{U}=(h,u,v)^T \tag{2.13}$$

$$
\mathbf{F} = \begin{pmatrix} hu \\ gh + \dfrac{1}{2}u^2 \\ uv \end{pmatrix}, \quad
\mathbf{G} = \begin{pmatrix} hv \\ uv \\ gh + \dfrac{1}{2}v^2 \end{pmatrix}, \quad
\mathbf{R} = \begin{pmatrix} 0 \\ g\left(s_{bx} - s_{fx}\right) \\ g\left(s_{by} - s_{fy}\right) \end{pmatrix} \tag{2.14}
$$

These equations are of hyperbolic type; hence the equations have two sets of characteristics. Disturbances can travel both upstream and downstream depending on the Froude number.

1D models are often based on the full dynamic equations, e.g. HEC-RAS (Hicks and Peacock, 2005) or SWMM (Huber *et al.*, 1988). In contrast, only few 2D models apply these equations, e.g. TELEMAC-2D (Horritt and Bates, 2002) or DIVAST model (Lin *et al.*, 2006), the reason being that: some results show that neglecting the advection terms does not have a significant impact on model predictions. However Lin et al., (2006) showed that if both advective and local accelerations are neglected a slowdown on the wave propagation will become visible.

Non-convective acceleration wave models
These models include all the terms in the Saint Venant equation except the convective acceleration term. These models are misnamed as 'non-inertial'. They are also referred as gravity wave models. Such models have recently been used for urban flood modelling because two-dimensional flow over an inundated urban flood plain is assumed to be a slow, shallow phenomenon and therefore the convective acceleration terms can be assumed to be small compared with the other terms. The convective acceleration term is said to be responsible for the oscillatory behaviour of the equations' solution (Aronica and Lanza, 2005). The non-convective acceleration wave models, therefore, benefit from an increase of stability while keeping the ability to propagate along two characteristic directions.

Non-local acceleration wave models
In these models, the local acceleration is no longer considered. This approximation is preferable for flows where inertia terms are dominant. The equations are hyperbolic (Ponce, 1990). These equations were successfully applied in a 2D depth-averaged gradually varying free surface flow model (Leandro, 2008).

Diffusive model without inertia

In these models, both the local and convective accelerations are neglected. The equations are of parabolic type and only have one set of characteristics in the same direction as the fluid flow. This gives these models the advantage that the two set of equations can be reduced to a single equation, the convection-diffusion equation. Nonetheless, this model still produces wave diffusion as well as wave translation. The effect of diffusion is the mechanism that enables it to reproduce backwater effects (Ramos and Almeida, 1987). These models are often used in 2D overland flow modelling which are commonly described as urban inundation models (Hsu *et al.*, 2000) and in the modelling of the transport of pollutants (Abbott, 1980). If the sum of the local acceleration (a measure of unsteadiness) and the advective acceleration (a measure of nonuniformity) is small compared to the sum of the weight (i.e., gravitational) and pressure components, this model is capable of producing a simulation virtually as realistic as the dynamic wave model. This is often the case for flows at a low Froude number (US Army Corps of Engineers, 1993).

Kinematic wave model

If the pressure term and the acceleration terms become small compared to the bed and friction slope terms, such as in the case of relatively steep bed slope, and the change in discharge is moderate, the friction slope and the bed slope are approximately in balance. This is called the kinematic wave approximation, thus kinematic models consider only the gravity and friction terms. The flow is based on a friction law, and, the relationship between the depth and flow becomes unique. The set of equations can be reduced to a single equation resulting in only one set of characteristics. Since waves can only travel downstream, this model is inadequate for cases where backwater effects need to be considered. In this model waves translate at a constant speed (celerity) but do not suffer any attenuation (damping). Despite this, the model can exhibit an "artificial-diffusion" dependent on the numerical diffusion of the scheme, which is a function of both the space and time step used. This model is often used for flow routing in rivers (Price, 2009a; b) calculating runoff or initial conditions.

2.2.5   The Boussinesq equations

The Boussinesq equations describe flows for which the pressure distribution is not hydrostatic. The Boussinesq equations have terms accounting for the non-hydrostatic pressure distribution additional to those in the Saint Venant equations. If these additional terms are neglected, the Boussinesq equations reduce to the Saint Venant equations (Chaudhry, 1993). Looking at the assumptions used to derive the Saint Venant equations, it is readily seen that the equations are not valid if the pressure distribution is not hydrostatic such as when modelling steep wave fronts. If the flow streamlines have sharp curvatures, the pressure distribution can be no longer hydrostatic. In this case the Boussinesq equations (Chaudhry, 1993) should be used.

$$\frac{\partial(hu)}{\partial t}+\frac{\partial}{\partial x}\left(hu^2+\frac{1}{2}gh^2-\frac{h^3\partial^2 u}{3\partial x\partial t}-\frac{h^3 u\partial^2 u}{3\partial x^2}+\frac{h^3}{3}\left(\frac{\partial u}{\partial x}\right)^2\right)+g\left(s_{fx}-s_{bx}\right)=0 \tag{2.15}$$

2.2.6 The Pressurized flow

The continuity and momentum equations that describe transient-state flows in closed conduits are shown in (2.16) and (2.17). In these equations, there are two independent variables $(x,\ t)$ and two dependent variables: the piezometric head above a specified datum and flow (H,Q). Several simplifying assumption are made in the derivation of these equations (Chaudhry, 1987). In this case the following assumptions are taken into account:

1. The pressure distribution is hydrostatic (the flow streamlines do not have sharp curvatures).
2. The lateral inflow is nil.
3. The pipe has a linear elastic behaviour.
4. The flow velocity is uniform across the pipe cross section.
5. The flow is gradually varied.
6. The head loss is approximated by the steady-state resistance laws.
7. The liquid is incompressible and homogeneous.

Continuity equation

$$\frac{\partial H}{\partial t}+\frac{a^2}{A}\frac{\partial Q}{\partial x}=0 \tag{2.16}$$

Momentum equation

$$\frac{1}{A}\frac{\partial Q}{\partial t}+\frac{1}{A}\frac{\partial}{\partial x}\left(\frac{Q^2}{A}\right)+g\frac{\partial H}{\partial x}=g\left(S_b-S_f\right) \tag{2.17}$$

where a a is the celerity of the pressure wave which is defined as (Chaudhry, 1987);

$$a=\sqrt{\frac{K}{\rho\left[1+(K/E)\psi\right]}} \tag{2.18}$$

21

Here ψ is a nondimensional parameter that depends upon the elastic properties of the conduit; E is the Young's modulus of elasticity of the conduit walls; and K and ρ are the bulk modulus of elasticity and density of the fluid, respectively. Expressions for ψ for various elastic condition of conduits is given in (Chaudhry, 1987).

The friction term S_f in (2.17) is replaced by the Colebrook-White equation, valid for the transition zone from laminar to fully turbulent flow on the Moody diagram. Alternatively it can be approximated less accurately by the Hazen-Williams or the Chézy-Manning equation (Potter and Wiggert, 2002).

Although the equations for gradually varying free surface and pressurized flow are similar, they do not provide a unified system of equations for the mixed type of flow (free surface-pressurized flow). Therefore, the task of accurately simulating the transition state between free and pressurized flow becomes complex (Leandro, 2008).

2.3   Numerical Schemes

The conservation equations for mass and momentum are more complex than they appear. They are non-linear, coupled and difficult to solve. It is difficult to prove by the existing mathematical tools that a unique solution exists for particular boundary conditions. Experience shows that the Navier-Stocks equations describe the flow of a Newtonian fluid accurately. Only in a small number of cases - mostly fully developed flows in simple geometries such as in pipes, between parallel plates etc. - an analytical solution of the Navier-Stocks equations is possible (Ferziger and Peri 1999). Unsteady gradually varied flow is better described by the full dynamic equations, for both free-surface and pressurized flow. An analytical solution of these equations does not exist except for simplified cases. In most cases even the simplified equations cannot be solved analytically, therefore, one has to use numerical methods. To obtain an approximate solution numerically, we have to use a discritization method which approximates the differential equations by a system of algebraic equations, which can then be solved on a computer. The approximations are applied in space and/or time so that the numerical solution provides results at discrete locations in space and time. Numerical results are always approximate. Errors arise from each part of the process used to produce numerical solutions. The differential equations may contain approximations or idealizations; approximations are made in the discretization process and in solving the discretized equations; iterative methods are used and unless they are run for a very long time, the exact solution of the discretized equations is not produced. Each numerical method has its own advantages and disadvantages that, depending on the problem specifications,

22

influence the quality of the results. Problems such as numerical diffusion, oscillations and instabilities can be explained by an inadequate choice of parameters or/and numerical scheme.

Some of the numerical schemes often used in the literature, which are well established in the science community and applied in urban flood models are described briefly below, together with their advantage and disadvantages.

There are numerous numerical techniques available for solving problems based upon conservation laws. With the advances in computer technology various numerical methods have been developed and tested thoroughly. The Method of Characteristics (MoC), Finite-difference methods (FDM), Finite-element methods (FEM), and Finite Volume Methods (FVM) are among the available methods which are described in this section.

2.3.1 The Method of Characteristics

There are various references describing the MoC. The description presented here is mainly based on material taken from Cunge et al. (1980) and Crossely (1999).

The method of characteristics can only be applied to hyperbolic Partial Differential Equations (PDEs) and involves defining the characteristics along which disturbance propagate (Cunge *et al.*, 1980). Characteristics can be thought of as lines in a space-time plane, along which (by definition) certain properties are constant. The basis of the method can be illustrated considering a first order PDE of the form

$$\frac{\partial u}{\partial t} + a(x,t)\frac{\partial u}{\partial x} = 0 \tag{2.19}$$

with initial condition $u(x,0) = u_0(x)$.

By the chain rule;

$$\frac{du}{dt} = \frac{\partial u}{\partial t} + \frac{dx}{dt}\frac{\partial u}{\partial x} \tag{2.20}$$

and rearranging (2.20);

$$\frac{\partial u}{\partial t} = \frac{du}{dt} - \frac{dx}{dt}\frac{\partial u}{\partial x}$$

substituting the above expression in (2.19) yields

$$\frac{du}{dt} + \left(a(x,t) - \frac{dx}{dt} \right) \frac{\partial u}{\partial x} = 0 \tag{2.21}$$

From (2.21) it can be seen that $\frac{du}{dt} = 0$ along the lines defined by $\frac{dx}{dt} = a(x,t)$, which

implies that the solution u is constant along these lines known as the characteristics. In principle, if one can define a set of characteristics lines then it is possible to know the solutions at all times (providing the lines do not intercept) just from the initial and boundary conditions of the problem. Mathematically, this is equivalent to saying

$$u(x,t) = u\left(x - \int_0^t a(x,t)dt, 0 \right) \tag{2.22}$$

If the method is applied over a finite region, then it is necessary to specify the values along any boundary where the characteristics enter the region.

Hyperbolic PDEs admit discontinuous solutions, and for a general nonlinear conservation law with arbitrary initial conditions, the characteristics will cross in finite time and a discontinuity and shock will form. In this instance it is not possible to track back along the characteristic paths to find the solution. In free-surface flow the MOC fails because of the convergence of characteristic curves whenever a bore or shock forms and as such the method is not used as much in free-surface flow as it is in pressurized flow.

2.3.2   Finite Difference Method

The finite difference method is the oldest method for numerical solution of PDEs, believed to have been introduced by Euler in the 18th century (Ferziger and Peri 1999). A finite difference method represents a problem through a series of values at particular points or nodes. Expressions for the unknowns are derived by replacing the derivative terms in the model equations with truncated Taylor series expansions. The earliest numerical schemes are based upon finite difference construction and are conceptually and intuitively one of the easier methods to implement. However, fundamentally, such techniques require a high degree of regularity within the mesh and so this limits their application to complex problems. Also the disadvantage of these methods is that the conservation is not enforced unless special care is taken.

2.3.3   Finite Volume Method

The finite volume method uses the integral form of the conservation equations as its starting point. It is based upon forming a discretisation from an integral form of the model equations. The solution domain is subdivided into a finite number of contiguous control volumes (CV) and the conservation equations are applied to each CV. At the centroid of each CV lies a computation node at which the variable values are to be calculated. Interpolation is used to express variable values at the CV surface in the form of the nodal (CV-center values). With the emphasis of most fluid modelling problems being based upon conservation principles, the finite volume method has become the more popular approach for general fluid flow problem. The finite volume method can accommodate any type of grid, so it is suitable for complex geometries. The disadvantage of this method compared to the finite difference method is that methods of order higher than second are more difficult to develop in 3D. This is due to the fact that the finite volume method approach requires three levels of approximation: interpolation, differentiation and integration.

2.3.4   Finite Element Method

The basis of the finite element method is to divide the domain into a set of discrete volumes of finite elements that are generally unstructured; in 2D they are usually triangles or quadrilaterals, while in 3D tetrahedra or hexahedra are most often used and to place within each element nodes at which the numerical solution is determined. The distinguishing feature of finite element methods is that the equations are multiplied by a weighting factor before they are integrated over the entire domain. The solution is approximated by a shape function, within each element, in a way to assure continuity across each element's boundaries. This approximation is then substituted into the weighted-integral of the conservation-law and the equations to be solved are derived by imposing the derivative of the integral to be zero in respect to each nodal value.

An important advantage of the finite element method is its ability to deal with unstructured grids, and its main disadvantage is that the matrices are not well structured, which imposes difficulties on solution methods.

2.3.5   Explicit and Implicit Numerical Schemes

Numerical solution schemes are often referred to as being explicit or implicit. When a direct computation of the dependent variables can be made in terms of known quantities, the computation is said to be explicit. When the dependent variables are defined by coupled sets of equations, and either a matrix or iterative technique is needed to obtain the solution, the numerical method is said to be implicit. For example, taking the linear advection equation

$$\frac{\partial u}{\partial t} + a \frac{\partial u}{\partial x} = 0 \quad a = constant$$

then if the solution is to be advanced to time level $n+1$, the spatial derivative may be approximated either in terms of the known values at time level n or the unknown quantities at level $n+1$. If an approximation for the spatial derivative is approximated at time level n then that corresponds to an explicit method, whereas using level $n+1$ represents an implicit formulation. Both explicit and implicit schemes have their relative merits. Explicit schemes are conditionally stable under the Courant-Friedrichs-Lewy (CFL) condition which sets a limit on the maximum allowable time step and are generally simpler in terms of the resulting algebraic equations as implicit schemes usually require a matrix inversion which is more costly. However most implicit schemes are not restricted by the CFL stability constraints placed upon the explicit counterparts, and so allow the use of much larger time steps.

Although implicit schemes do not have to respect Courant's stability criteria (Chaudhry, 1987), they suffer from instability derived from oscillations whenever the changes of flow or depth are too rapid within a time step. This is more likely to occur in the presence of small water depths; in this case, depth oscillations can be sufficient to cause the simulation to fail.

2.3.6 Numerical Grid

The discrete locations at which the variables are to be calculated are defined by the numerical grid which is essentially a discrete representation of the geometric domain on which the problem is to be solved. It divides the solution domain in to a finite number of subdomains. Early efforts focused on using regular or structured grids, where by all the cells or elements were of the same size. Regular grid is the simplest grid structure, since it is logically equivalent to a Cartesian grid. Although this has advantage in terms of numbering the cells and forming the discrete equations which simplifies programming and developing a solution technique, they can be used only for geometrically simple solution domains. Another disadvantage is that it is difficult to control the distribution of grid points: concentration of points in one region for reasons of accuracy produces unnecessarily small spacing in other parts of the solution domain and leads to unnecessary computational expense. Unstructured or irregular grids overcome the difficulties associated with regular grids. Irregular grids are the most flexible type of grids which can fit very complex geometries. However, the resulting meshes generally have no apparent structure and therefore increase the level of complexity of generating suitable computer code. The predominant reason for using irregular gridding is the ability to concentrate the cells in areas where sharp gradients occur, and so high level of accuracy can be maintained throughout the solution domain without the need to use a fine grid everywhere.

Finite difference schemes used principally for the solution of gradually varying free surface flow equations utilise either staggered or non-staggered discretisation grids. Non-staggered grids have the two dependent variables placed at the same discretisation points while staggered grids alternate the dependent variables. The most widely used numerical schemes are the Preissmann 4-point Implicit scheme on non-staggered grids (Liggett and Cunge, 1975) and Abbott-Ionescu 6-point Implicit scheme on staggered grids (Abbott and Ionescu, 1967; Cunge et al., 1980). Both of these schemes offer a linearised discretisation of the governing equations resulting in a system of linear simultaneous equations to be solved for each time step. In the case of the Abbott-Ionescu scheme the system matrix is tridiagonal while in the case of Preissmann scheme the system matrix is pentadiagonal. When accompanied by proper boundary conditions these systems of equations are usually solved by a direct, efficient method called the double sweep. It is a form of Gaussian elimination for banded matrices. The usual requirement for the system matrix to be solvable by this method is that it is diagonally dominant. The system of equations obtained by the finite difference approximation has two unknowns more than the number of equations. Adding two equations representing the boundary conditions solves this problem.

Amongst the two implicit finite difference schemes, Preissmann scheme has advantages over Abbott-Ionescu scheme since it allows non-equidistant grids and computes dependant variables at the same point. Meselhe and Holly (1997) showed that it cannot be used to simulate transcritical flow and Sart et al, (2010) presented adaptation of the Preissmann scheme for transcritical open channel flows.

Another implicit scheme which is used in solutions of gradually varying free surface flow equations is the Alternating Direction Implicit (ADI) schemes. This method is used by (Peaceman and Rachford, 1955) to numerically approximate the solution of heat flow in 2D. According to (Peaceman and Rachford, 1955) the method provides greater superiority to the explicit finite difference method due to the high computational efficiency which requires less computing time because it involves a tridiagonal matrix.

The explicit difference equation is simple to solve but it requires an uneconomically large number of time steps-of undefined size and the full implicit difference equations do not limit the time step but they require a complex matrix inversion that is very time consuming at each time step for the solution of large set of simultaneous equation. The ADI method has subsquently been used and improved by others such as Leendertse (1967), Stelling et al (1986) and Falconer (1980).

In the ADI algorithm the solution procedure is split in such a way that in one direction the conservation of mass and conservation of the direction-corresponding momentum are solved, and, after that, in the other direction, the conservation of mass is again solved but now with the conservation of momentum introduced in that direction. For this purpose, the time step is divided into two parts and the equations are solved sequentially in the x and y directions in two half time steps $t^{n+1/2}$ and t^{n+1} respectively.

Roe and MacCormack schemes are also widely used in solutions of gradually varying free surface flow equations. Roe scheme is a numerical method used to solve Riemann problems. A Riemann problem is one where the initial data is constant either side of a discontinuity. Roe scheme is an upwind scheme which is based on the idea of discretizing the special derivatives so that information is taken from the side it comes. It has been proved that Roe scheme can be applied to the non-linear hyperbolic shallow water equations; see (Alcrudo *et al.*, 1992; Crossely, 1999; Glaister, 1988).

MacCormack introduced a two-stage numerical scheme for compressible flows with a predictor stage followed by a corrector stage. The MacCormack scheme is a fractional-step method where a complicated finite-difference operator is 'split' into a sequence of simpler ones. The splitting process reduces the number of calculations during each time step and achieves second-order accuracy in space and time when a symmetric sequence of operators is used. The scheme is widely applicable, in part, because of its simplicity and robustness see (Garcia-Navarro and Brufau, 2006; Garcia and Kahawita, 1986)

2.3.7   Accuracy, consistency, stability, convergence and well posedness

In order to quantify how well a particular numerical technique performs in generating a solution to a problem, there are four fundamental criteria that can be applied. The four criteria are accuracy, consistency, stability and convergence. In theory these criteria apply to any form of numerical method though they are most easily formulated for finite difference schemes. The following is based upon descriptions from (Smith, 1985) and (Versteeg and Malalasekera, 2007).

Accuracy is a measure of how well the discrete solution represents the exact solution of the problem. Two quantities exist to measure this: the local or truncation error, which measures how well the difference equations match the differential equations, and the global error which reflects the overall error in the solution and in reality is not possible to find unless the exact solution is known. An expression for the truncation error can be obtained by substituting the known exact solution of the problem into the discretisation, leaving a remainder which is then a measure of the error. Alternatively, the exact solution to the discretised problem could be substituted into the differential equation and the remainder obtained. For example for a PDE this would lead to an expression of the form $\tau = o(\Delta t^q, \Delta x^p)$ where τ is the truncation error and Δt and Δx are the time and spatial steps (assuming a regular grid). From this, the method is said to be q^{th} order in time and p^{th} order in space, and generally this is referred to as the level of accuracy of the scheme. It is natural to assume that by increasing the grid resolution then any errors will be reduced and this leads to the definition of consistency. Mathematically, for a method to be consistent then the truncation error must decrease as the step size is reduced, which is the case when $q.p \geq 1$, which is equivalent to saying that as $\Delta t, \Delta x$ tend to zero then the discretised equations should tend towards the differential equation. For a scheme to be of practical use, it must be consistent.

Formally if a scheme is said to be stable, then any errors in the solution will remain bounded. In practice if an unstable method is used then the solution will tends towards infinity. Most methods have stability limits which place restrictions on the size of the grid spacings (i.e. $\Delta t, \Delta x$) that can be used, usually in terms of a limit on the CFL (Courant-Friedrichs-Lewy) number. A number of methods are available for obtaining expressions for the stability conditions, and the appropriate choice depends on the actual problem.

Another requirement is that the numerical scheme should be convergent, which by definition means that the numerical solution should approach the exact solution as the grid spacing tends to zero. This is coupled with the global error. However it is usually not possible to prove the convergence of a particular scheme to a specific problem. Instead use is made of Lax's Equivalence theorem which states that for a well posed initial value problem (IVP) and a consistent method, stability implies convergence, in the case of a linear problem. For non-linear equations, stability and consistency are necessary but not sufficient conditions for convergence. These criteria dictate whether a particular numerical scheme is suited to solving a particular problem. There is another condition, which has to be satisfied in order to produce a valid solution and this relates to the actual problem and is the issue of well posedness. In order to generate a numerical solution, the problem being considered must be well posed. For this to be the case then the following conditions must hold;

  a) a solution must exist
  b) the solution should be unique
  c) the solution should depend linearly on the data in some way.

The last condition can be translated to mean that the solution should not be sensitive to small changes in the initial/boundary data of the problem. If a problem is not well posed, then a valid numerical solution cannot be generated and any numerical treatment will either fail or produce poor results. One easy way to produce an ill posed problem is to apply inappropriate boundary conditions, for example by trying to enforce values of the quantities being modelled on the characteristics leaving the computational domain. If the initial data is not fully specified, then this also presents an ill posed problem as there will be no unique solution.

2.4   Urban Flooding

Urban areas can be flooded by rivers, coastal floods, pluvial and ground water floods, and artificial system failures. According to Jha et al., (2012), urban floods typically stem from a complex combination of causes, resulting from a combination of meteorological and hydrological extremes, such as extreme precipitation and flows. However they also frequently occur as a result of human activities, including unplanned growth and development in floodplains, or from the breach of a dam or an embankment that has failed to protect planned developments. Urban flooding occurs where there has been development within stream floodplains. Urbanization increases the magnitude and frequency of floods by increasing impermeable surfaces, increasing the speed of drainage collection, reducing the carrying capacity of the land and, occasionally, overwhelming sewer systems. High intensity rainfall can cause flooding when drainage systems do not have the necessary capacity to cope with flows. Sometimes the water enters the drainage system in one place and resurfaces in others. River floods can be slow, for example due to sustained rainfall, or fast, for instance as a result of rapid snowmelt. Floods can be caused by heavy rains from monsoons, hurricanes or tropical depressions. They can also be related to drainage obstructions due to landslides, ice or debris that can cause floods upstream from the obstruction Overland floods caused by rainfall or snowmelt that is not absorbed into the land flows over land and through urban areas before it reaches drainage systems or watercourses. This kind of flooding often occurs in urban areas as the lack of permeability of the land surface means that rainfall cannot be absorbed rapidly enough.

Flash floods occur within a few minutes or hours of heavy amounts of rainfall, from a dam or levee failure, or from a sudden release of water held by an ice jam. Although flash flooding occurs often along mountain streams, it is also common in urban areas. Urban areas are notably susceptible to flash floods because a high percentage of their surfaces are composed of impervious streets, roofs, and car parking areas where runoff occurs very rapidly (Jha *et al.*, 2012).

The primary cause of urban flooding, according to Andjelkovic (2001), is a severe thunderstorm or a rainstorm proceeded by a long-lasting moderate rainfall that saturates the soil. Floods in urban conditions are flashy in nature and occur both on urbanised surfaces (streets, parking lots, yards, parks) and in small urban creeks that deliver water to large water bodies.

Other causes of urban floods are (Campana and Tucci, 2001; Kolsky and Butler, 2002) :

- inadequate land use and channelisation of natural waterways
- failure of the city protection dikes
- inflow from the river during high stages into urban drainage system
- surcharge due to blockage of drains and street inlets
- soil erosion generating material that clogs drainage system and inlets
- inadequate street cleaning practice that clogs street inlets

Flooding in urban drainage system may occur at different stages of hydraulic surcharge depending on the drainage system (separate or combined), general drainage characteristics as well as specific local constraints. Urbanization causes change of the land characteristics from pervious to imperviousness. Asphalt and concrete, and rooftops replace forest trees and soil. Storm-water sewers replace stream channels. All increase runoff, and the important flood peaks that cause stream channel erosion and destruction of channels, property, and lives. Impervious surfaces that cover most of the urban area decrease the ability of the land to absorb rainfall-runoff and causes rapid surface runoff. The runoff from the increased pavement goes into storm sewers, which then goes into streams. This runoff, which used to soak into the ground, now goes into streams, causing flooding. Blockage within a drainage system, inadequate capacity of drains and heavy precipitation can sometimes be the main causes or urban flooding. Such flooding may cause large damage to residential and commercial buildings and public and private infrastructure.

2.5   Urban Flood Models

There is nowadays a range of modelling approaches available for modelling urban floods. Existing urban flood models greatly vary depending on their purpose, the governing equations solved, the numerical scheme used and dimensionality (1D, 2D and even 3D). Some models are developed for research purpose and some models are developed as a commercial software package. The commercial flood models are characterized by the use of similar, well established governing equations, and comprehensively tested numerical schemes and tend to be more robust, whereas flood models which are developed for research based applications in the scientific community are more heterogeneous and less robust compared to the commercial ones (Leandro, 2008).

In this section a review of some of the existing urban flood models and their modelling approach is presented. Part of this review is taken from Leandro (2008).

1D models

1D models are used to simulate flow through pipes, channels, culverts and other defined geometries. The system of 1D cross-sectional-averaged Saint-Venant equations, which are used to describe the evolution of the water depth and either the discharge or the mean flow velocity consists of conservation of mass (continuity equation) and momentum. The boundary conditions are discharges, water levels (or depths) or free flow conditions at conduit/channel ends.

In a 1D model, the network is composed of pipes and manholes. Pipes or open channels are represented by links while manholes are nodes that can also represent basins and outlets.

The 1D modelling approaches are usually limited to circumstances where overland flows are confined to predetermined flow paths. In this approach, if the water level in a manhole or basin reaches the ground level an artificial "inundation" basin is inserted above a node such as in MOUSE (developed by the Danish Institute of Water and Environment (DHI Group)) for surface run off, open channel flow, pipe flow, water quality and sediment transport for urban drainage system, storm water and sanitary sewers (DHI Group, 2009a). The surface area of this basin is gradually increased from the area in the manhole or the basin to a number of times larger area, thus simulating the surface inundation. When the outflow from the node is negative, the water stored in the inundation basin re-enters the system.

SWWM is the EPA Storm Water Management tool, which was first developed in 1971, and it has undergone several upgrades, until its latest version, EPA-SWWM5. In this version, flooding is a special case of surcharge which takes place when the hydraulic grade line breaks the ground surface and water is lost from the sewer node to the aboveground system (Rossman, 2006).

DRAINS is a Stormwater Drainage System design and analysis program which has been widely used for urban stormwater system design and analysis in Australia and New Zealand. O'Loughlin and Seneviratne (2006) analysed the DRAINS model and found that for complex cases of overland flow, the DRAINS model was not realistic and suggested that a surface model should be run separately either by using HEC-RAS or 2D models. In setting up the model, the author found it difficult to define accurately surface when a subsurface path is encountered and in cases where the flow overtops the pathways. The study concluded with the need for developing and integrated sewer and overland flow model.

1D-1D models

Approaches to modelling of urban flooding evolved in recent years with a significant step forward with the dual drainage concept (Djordjević et al., 2005), which consider the interactions between a minor system and a major system. The major system includes all above ground flood pathways (natural and man-made), including both open and culverted watercourses. The minor system is the drainage sewer network, including the manholes and the inlet connections. These systems are linked via weir or orifice-type elements representing inlets like catch pits and holes on manhole covers through which direct interaction between the two systems take place. This approach allows for dynamic simulation of flood flow movement with the results in the form of hydrographs local flood flow depths and velocities that can be used for analysing different flood mitigation schemes, damage evaluation and flood risk mapping.

The 1D-1D coupled hydraulic model solves simultaneously the continuity equation for the network nodes, the complete Saint Venant equations for the 1D sewers and the 1D above ground networks, and the links equations.

Mark et al. (2004) outlined the potential and limitations of a 1D-1D modelling technique where a flood model was built in two layers describing the conditions both in the surcharged pipe system and flooding on the catchment surface. The model incorporates the interaction between the buried pipe system, the streets as open channel flow and the areas flooded with stagnant water. The authors stated that the modelling approach was generic in the sense that it handled both urban flooding with and without flood water entry into houses. With regard to the surface network, it was found necessary to compute accurately storage level curves, and determine the flood pathways based on a GIS procedure using a Digital-Elevation- Model (Bolle *et al.*, 2006). The study concluded that during heavy flooding the 1D approach may be insufficient, and indicated the need to use a full 2D dynamic model to describe the surface flow.

Schmitt et al (2004) used the dual drainage concept to develop the dual drainage model called RisUrSim in order to meet the requirements of simulating urban flooding, focusing on the occurrence of distinct surface flow and its possible interaction with the surcharged sewer system. They used a uni-directional link between a hydrologic surface runoff simulation model and sewer flow routing model and a bi-directional link between a hydraulic surface flow simulation model and sewer flow routing model. The dual drainage concept was also applied by Vaes et al. (2004), when the authors used 1D-1D INFOWORKS CS (Wallingford-Software now MWHSoft, 2004).

Spry and Zhang (2006) used the 1D-1D models XP-SWMM and DRAINS to model two distinct case studies. The surface network was delineated based on digital-terrain-model (DTM), and fine-tuned based on cadastre, site reconnaissance and aerial photography. It was assumed that generally the surface pathways would follow kerb lines along driveways, roads and property boundaries. The study concluded that it was necessary to use integrated models, and that 1D-1D models can be an economic and robust alternative for 1D-2D models in case of lack of data or non existence of flat areas.

1D-2D models
Chen et al (2007) linked two research based applications, SIPSON with UIM. SIPSON solves simultaneously the continuity equations for network nodes, the complete Saint Venant equations for the 1D networks and the links equations (Djordjevic et al., 2005). The UIM is a 2D diffusive overland-flow model that solves the non-inertia flow equations. The proposed model, SIPSON/UIM, was said to simulate the complex flow process on overland-surface of pluvial flooding cases, more effectively than the 1D model.

The simulation process in the case of coupled 1D-2D modelling is based on complex numerical solution schemes for computation of water levels, discharges and velocities. 1D-2D coupled models use 1D unsteady flow calculations to simulate flow in pipes, channels, culverts and other defined geometries, and the 2D models are used where the flow is truly two dimensional. The 2D surface model simulates, for a give computational

domain, boundary conditions and an initial condition, the free water surface elevation and
the depth-averaged velocity vector. . The quality of the results will depend on one hand
on the quality of the numerical scheme, on the other hand on the quality and accuracy of
the data. Accurate description of features and structures on the surface such as rivers,
roads, levees, shores and details of buildings and small scale features such as curbs, small
hills and depressions and the inability to precisely determine the values of certain
parameters such as friction factor are major constraint in hydraulic modelling of such
models.

The two domains (1D and 2D) are normally coupled at grid cells overlying the channel
computational points through mutual points of the connected cell and the adjoining
channel section (Price and Vojinovic, 2010; Vojinovic and van Teeffelen, 2007). The
interactions between channels and floodplains are determined according to the type of
link between them.

Lhomme et al. (2006) compared a 1D GIS based model (kinematic) with the 2D model
RUBAR 20 (shallow water Equations). The authors concluded that the former model
accurately models the steepest streets but is less suitable for flatter streets. Moreover the
distribution of flows at crossroads and disregarding of backwater effects were expected to
be a major source of errors in the 1D model.

Phillips et al (2005) used XP-SWMM2D to simulate Cumberland highway sub-
catchments, Cabramatta (Australia). XP-SWMM2D is an integrated model that links 1D
XP-SWMM, to model the minor system, with the 2D TUFLOW, to model the major
system. The increased availability of aerial laser scanning (ALS) is said to provide
enough detail to support the 2D modelling. A flood extent comparison was done using
TUFLOW alone, and integrated with XP-SWWM. Strong differences were highlighted in
this comparison, encouraging the use of integrated linking. In this same study, a coupled
model was used to perform a review of flood levels in Prospect Creek floodplain. In this
case the 2D model was compared with the 1D-2D model, where the river was modelled as
a 1D channel. The 1D-2D was said to outperform the 2D based on loss of definition of
channel sections (using a 10 m grid). Having a 1D channel avoided the need for a finer
grid to be used on the river bed. TUFLOW was first released in 1990 as a joint project
from WBM Pty Ltd and The University of Queensland. The 2D solution is based on the
Stelling finite difference, alternating direction implicit (ADI) method. The 1D solution is
based on a two-stage explicit method. XP-SWWM is a software package copyright of XP
Software Inc. which uses SWWM as the hydraulic engine.

Bolle et al, (2006) studied the bi-directional interaction between the sewer network of Erpe-network in Belgium and the Molenbeeek River using the SOBEK model. Although SOBEK can also accommodate 2D simulations, both sewer and river were modelled using 1D models. This study indicates that whether the simulation objectives are upon the sewer or the river results, the bi-directional interaction becomes respectively more or less relevant This is to say that bi-directionality is required whenever the interests in sewer behaviour becomes gradually more important. Spatially the sewer water level differences become largely higher at the downstream ends, and decrease towards the upstream ends. The authors however concluded that this last statement is strongly system dependent. SOBEK is copyright of Deltares, Netherlands. It is based on the 1D full dynamic equations and the 2D shallow water equations, using an implicit scheme known as the Delft Scheme.

Carr and Smith (2006) presented a 1D-2D model which linked the 1D sewer network model called MIKE STORM with the overland 2D flow model MIKE 21 (DHI Group, 2009b). In this study, the performance of the 1D-2D model was compared to the 1D-1D MIKE STORM. The comparison was done based on a time series downstream-discharge on two catchments. The authors concluded that the 1D-2D model provided better fit to the time series, and it had the advantage of not having to delineate the flow paths, as compared with the 1D-1D model. MIKE STORM shares the hydraulic engine with MOUSE (DHI Group, 2009a), the first hydraulic model developed by DHI in 1983. This model is based on an implicit FDM.

2.6   Linking Surface with Subsurface Networks

An integrated model is defined as: simultaneously linking and simulating two or more systems (e.g. minor, major and underground systems). A coupled model is defined as: simultaneously linking and simulating two systems (e.g. minor and major systems).

In integrated or coupled urban flood models, the linking element defines the interacting discharge between the surface and subsurface system. The linking elements are the main features responsible for regulating the interacting discharged and thereby defining and constraining the extent of the flood inundation. The geometric characteristics of these linking elements play a significant role in the models ability to realistically model urban flooding. Therefore understanding of the physics and the governing equations which are used to calculate the interacting discharges through these linking elements is very important. The most common way in which the interacting discharges through the linking elements have been modelled is through the use of either a weir equation or an orifice equation, or a combination of both (Chen et al., 2007).

36

The Fundamental Equations used to calculate the discharge through weirs and orifices found in literatures and are briefly reviewed as follows. Four different types of weir/orifice equations can be found in the literature. These are Broad-Crested weir, sharp-crested, orifice and submerged wire. All can be derived using the Bernoulli and continuity equations.

1. **Broad crested weir**: A broad-crested weir is a flat-crested structure with a crest length large compared with the flow depth. In the broad-crested weir the streamlines become parallel, resulting in an hydrostatic pressure distribution. The flow equation can be obtained by applying the Bernoulli equation in a horizontal rectangular channel between upstream of the weir and on the crest and by neglecting the kinetic energy upstream from the broad weir.
   The discharge above the weir is given by;

$$Q = \frac{2}{3} b \sqrt{\frac{2}{3} gH^3}$$ (2.23)

   where H is the upstream head above the crest level and b is the channel width.

2. **Sharp-crested weir**: A sharp-crested weir is characterized by a thin sharp-crest. In the absence of sidewall contraction, the flow is basically two dimensional and the flow field can be solved by analytical and graphical methods (Chanson, 2004). For aerated nappe, the discharge is expressed as;

$$Q = Cd \frac{2}{3} b \sqrt{2gH^3}$$ (2.24)

   where Cd is a dimensionless discharge coefficient of a sharp-crested weir.

3. **Orifice**: Hydraulic analysis of an orifice is a simple application of Bernoulli equation between two points, before and after the orifice, along the centre streamline, in a horizontal duct with area A, which has an orifice plate with area A_0. The flow rate is expressed as;

$$Q = CdA_0 \sqrt{2gH}$$ (2.25)

   where H is the difference of the upstream and downstream waterlevels and Cd is given as a function of the reduction of area by the orifice plate, A to A_0, the contraction coefficient due to the "vena contracta" downstream of the orifice, and another coefficient to account for the unknown exact values at the section area of the "vena contracta".

4. **Submerged Weir**: A **transition** from a free overfall to a submerged flow passes through a jump-wave and follows when the latter disappears, which is a consequence of its submergence in the downstream side (Nikolov *et al.*, 1978). There are different equations proposed for the submerged weir. Villmonte (1947) applied the principle of superposition to the submerged linear weir flow problem. He assumed that flow over a submerged linear weir was equal to the difference between the free-flow discharge associated with the upstream water depth above the crest level and a free-flow discharge associated with the downstream depth water depth above the crest level. He proposed a reduction factor for the discharge as follows;

$$Q_s = Cd_f Q_1 \tag{2.26}$$

where Q_s is the submerged weir flow, Q_1 is the free weir flow given by (2.23) and C_{df} is the discharge reduction factor give as

$$Cd_f = \left(1 - \frac{H_2^{3/2}}{H_1^{3/2}}\right)^{0.385} \tag{2.27}$$

where H_2 and H_1 are the piezometric heads downstream and upstream respectively. The exponent of 0.385 (2.27) was determined based on the experimental result of Villemonte's submergence testing with seven different sharp-crested linear weir geometries.

Later in 1952, Gibson assumed that the discharge could be broken into two parts. The upper part of depth H_2- H_1, is defined as a weir equation, and the lower part of depth H_2, is defined as a submerged orifice (Leandro, 2008).

$$Q = Cd_4 \frac{2}{3}\sqrt{2gb}(H_1 - H_2)^{3/2} + Cd_5 \frac{2}{3}\sqrt{2gb}H_2(H_1 - H_2)^{1/2} \tag{2.28}$$

where Cd_4 and Cd_5 are the discharge coefficients of the weir and the orifice part.

Cunge et al. (1980) suggested replacing $Cd\frac{2}{3}$ in (2.24) by CK_w and the hydraulic gradient by the difference in both heads.

$$Q = CK_w H_2 \sqrt{2g} b \ (H_1 - H_2)^{1/2}$$ (2.29)

2.7   Topographic Data for Urban Flood Modelling

Topographic data are one of the most important data items required to build flood inundation models. Topographic data play significant role in urban flood modelling as they provide the modeller with land surface and topography representation that drives surface flow and is one of the most important data sources for deriving variables used in urban flood modelling (Wechsler, 2007). Though use of topographic data in hydrologic studies is ubiquitous, until relatively recently, topographic data were generated from contour lines produced by interpolating spot heights measured by hand on site. This is expensive and time consuming (Wright et al., 2008) and leads to coarse approximations of the overall surface and low accuracy with poor spatial resolution. Rapid technological advances, however, enabled the generation of highly detailed topographical maps over large areas in a relatively short amount of time and answered the need for high resolution spatially distributed data for model construction and validation. Topographic data produced from technologies such as Light Detection and Ranging (LiDAR) and IFSAR (Interferometric Synthetic Aperture Radar sensor) are more readily available (Wechsler, 2007). Remote sensing data taken through the use of LiDAR (Light Detection and Ranging), a modern technology based on the combination of a laser scanner, a Global Positioning System and an Inertial Measurement Unit, mounted on an aircraft, are mostly used to give a comprehensive topographic coverage of the entire area in an accurate and economic manner (Vojinovic and van Teeffelen, 2007). LiDAR data points are provided at a typical density of up to several dozens per square meter, and accuracy of better than 0.15 m vertically and 0.50 m horizontally. Once turned into grids at resolution 1~2 m, the data are processed to identify and/or remove surface features such as buildings, vegetation, etc (Abdullah et al., 2011; Néelz and Pender, 2008). High resolution DEM that can be gathered from LiDAR is proved to have a capability to solve problem associated with important small scale features and inadequate representation of topographical features and is rapidly gaining acceptance as a tool to generate necessary data for urban flood modelling work.

A DTM that includes surface features such as buildings is customarily referred to as a Digital Elevation Model (DEM). The most widely used DTMs and DEMs for flood modelling applications are made of a collection of surface elevation values on a regular square grid. According to Néelz and Pender (2008), this format is often favored by remote sensing data providers because it facilitates processing (compared to the unstructured Triangular Irregular Networks).

Ideally, detailed prediction of flood flows over complex topographies and in urban areas requires hydraulic model simulations on high resolution grids that can resolve the effects of individual buildings and other topographic features that play significant roles in routing and storing surface flow and are likely to have an influence on the movement of water across the flood plain. These could be natural pits, ridges, channels and man-made features. Resolving surface water movement through urban areas requires resolution of complex flow paths around buildings, representation of microscale topographic and blockage effects and numerical schemes capable of dealing with high-velocity flow at shallow depth. This requires high resolution model grids of the order of 1-5m (Hunter *et al.*, 2008). However, the associated computational cost due to smaller computational time steps may make this approach unfeasible where there is a need to simulate many flood event realisations (strategic risk planning), where modellers are encouraged to run multi-simulations in order to assess the sensitivity of model results to parameter and input data uncertainties and/or where there is limited time or resources available to perform the computer simulations (real-time forecasting, emergency planning, studies with limited availability of staff or computer resources). In such circumstances, topographic data is often generalised to a more manageable resolution and floodplain models are built at much coarser resolutions such that complicated flow patterns due to vegetation, buildings or any other man-made structures become sub-grid scale processes. Such models are often used to predict the general trends on the flood inundation. Such coarse grid models are informative, although to a lesser extent than fine grid models. There is therefore a trade-off between predictive capability and computational efficiency which needs to be addressed in a different manner depending on the intended application of the model (Néelz and Pender, 2008).

It is now widely recognised that topographically, model results and accuracy are sensitive to the resolution of the source data. As the spacing of elevation samples increases, land features and shapes that may have a significant impact on flood flow routes may be too fine to be represented accurately enough and the surface becomes more generalised and as a consequence, complicated flow patterns due to vegetation, buildings or any other man-made structures cannot be modelled.

2.7.1   Resolution scale

The choice of required resolution varies from application to application and it depends on many factors including the size of the study area, the minimum size of key features required by the model, the complexity of model and, the type of model used for simulation. For example, when modelling the flow of water on a large scale like that in terms of examining outflows from an upstream catchment into a river then a coarse resolution would be used, such as 100m^2 cells as used by (Jain and Singh, 2005), as at this scale the effects of buildings need not be considered. For 2D surface flow simulation within a smaller urban environment, a high resolution DTM is required in order to keep key information such as building and road features.

Researchers tried to find optimum resolution of DEM mathematically for a given application. Zhilin (2008) used a method based on local variance. The local variance is defined as the average of the variances within a moving window passing over the entire area with the DEM. The resolution with the maximum variance is then regarded as optimum.

2.7.2   Coarse resolution DTM

Generalization of topographic data can cause alteration or loss of important feature types. The alteration or loss of these features can pose problems within the hydrological modelling as key features like ponds may become shallower, channels may become wider, and buildings may spread and take up larger areas within the landscape. Within urban flood modelling it is the man-made building features that play a significant role in the routing of surface flow at and it is these key features that are distorted or lost during the generalisation process. With these changes within the landscape during generalisation 2D models with a lower resolution are more likely to result in flooding over a wider area and at shallower depths than high resolution 2D models.

According to Wood (1996) topographical features can generally be classified into 6 simple specific types as plane, channel, ridge, pass, peak and pit and the generalisation process can result in the subsequent changing of these features.

The generalisation of DSMs within urban environments leads to significant changes in the topology due to the spreading or loss of dominant features. To avoid the errors generated, it is common practice to remove the buildings from the model and interpolate ground heights in their location prior to generalisation. Yu and Lane (2006 (a)) stated that, the presence of significant features (houses, walls, etc.) on a river floodplain is important in both the volume of the floodplain that can be occupied by the flow and the direction that the flow takes across the floodplain. This fact is also true for pluvial flooding; therefore the building information needs to be maintained to some degree as to ensure the accuracy of the derived floodplain for a flood event.

To model urban areas using coarse grids more accurately, information is still required to represent fine detailed features present within each coarse grid cell such as building, walls, and fences. To reduce inaccurate flood model simulations due to the loss of features in the generalisation process methods were devised in order to keep the information that can be obtained from high resolution topographic data.

One of the methods to solve this problem was a multi-scaled approach which enabled the representation of terrain in various resolutions simultaneously. In regions where there are no significant fine resolution objects present the data can be represented by coarse grid-cells and where fine detail objects are present finer resolution grid-cells can be used (Evans, 2010). The key issue confronted with using a multi-scaled approach is that the data-set will be in an irregular grid format which is computationally and mathematically more complex to incorporate into models as opposed to regular-grid representations.

Regular grid representations are often the preferred choice of data format due to their ease in manipulation and representation within GIS packages. There are a large variety of methods available for generalising regular grids and the easiest way of generalisation (besides re-sampling), is a low-pass filter (Burrough *et al.*, 1998). This is a global filtering technique which does not distinguish between geomorphologic characteristic and uncharacteristic features.

A building's influence within a generalised grid-cell can be referenced separately to some degree numerically, by increasing the Manning's roughness coefficient n within the coarse grid-cell. The up-scaling of n however may reduce fluxes across linear sets of grid cells, but will not necessarily recognise the full topographical nature of the structure (Yu and Lane, 2006 (a)).

2.8   Rainfall Intensity and Infiltration Rate in Urban Flood Modelling

In 2D flood modelling rainfall intensity and infiltration rate are included as source and sink terms. 2D shallow water models are suitable to include spatially variable rainfall and soil infiltration due to the discrete representation of the geometric domain on which the equations are to be solved. 2D models which are capable of considering the spatial variability of rainfall, soil characteristics and land-use allow simulation of overland flow and the infiltration process during complex storm events recorded by multiple rain gauges. However, many previous studies of modelling 2D overland flow (Hromadka II *et al.*, 1987; Tayfur *et al.*, 1993; Zhang and Cundy, 1989) have limited their application to constant rainfall intensity and spatially constant infiltration rate. Esteves et al. (2000)

considered a time varying rainfall intensity, but the rainfall intensity and infiltration were spatially constant.

The infiltration rate is usually calculated using the Green-Ampt method (Green and Ampt, 1911) or a modified Horton equation for infiltration (Bauer, 1974). Description of the modified Horton equation for infiltration is presented here mainly based on material taken from (Bauer, 1974).

Proposed in 1939, Horton's (1939) widely accepted infiltration equation takes the form:

$$f = f_c + (f_o - f_c)e^{-kt} \qquad (2.30)$$

where
f_o is the initial infiltration capacity (mm/h);
f_c is the final or equilibrium infiltration capacity (mm/h);
f is the infiltration capacity (mm/h) at time t (h);
k is an exponent governing the rate of decline of infiltration capacity (1/h).

The equation represents a family of curves fitted to experimental results derived from the observed behaviour of natural soils and accordingly the influence on infiltration of soil texture and structure, root system development, earthworm perforations, etc. is taken into account in evaluating the parameters.

As the infiltration capacity f is expressed as a function of time rather than of soil water content, the equation holds true only if throughout the period of simulation rainfall intensity is greater than f. Another drawback is that variations in antecedent conditions cannot readily be taken into account.

The rate of infiltration depends on a wide variety of factors, such as soil water storage, crumb structure, porosity, etc. As infiltration continues, soil water content increases, earthworm perforations and cracks become clogged, initial root zone moisture deficit becomes satisfied, soil texture and structure may change, with the inevitable result that the rate of infiltration declines with input.

Bauer (1974) attempted to account for changes in infiltration capacity resulting from the combined effects of all the temporal changes in state, the most important of which is likely to be the change in soil water content. The term soil water content is accordingly assumed to represent all the factors that influence the potential rate of infiltration.

Following this concept (Bauer, 1974) assumed that a highly compacted soil having a low soil water storage displays the same infiltration capacity as a highly porous soil having a high soil water content. Consequently, such soils will have the same 'equivalent soil water content'.

In Horton's equation, infiltration capacity f is a function of time t and so, if equation (1) is employed in a simulation programme, f can but decrease as simulation progresses, whereas realistically it should be possible for f either to decrease or to increase. If the potential infiltration rate is at all times equalled or exceeded, f will decrease continuously, but should the rainfall rate drop below the potential infiltration rate, f should either remain constant or rise, depending upon the input and drainage rates.

According Bauer (1974), realistic behaviour can be simulated by taking into account the soil water content. At low water content the potential rate of infiltration will be higher than when the soil is wet. As the soil wets up the rate of infiltration declines and the rate of percolation or drainage rises. Maximum infiltration capacity corresponds to minimum soil water content and vice versa. Alternatively, minimum drainage rate corresponds to minimum soil water content and vice versa. Finally, for continuity, minimum infiltration rate approximates to maximum drainage rate.

Bauer (1974) incorporated the above concept in Horton's equation through the introduction of a function to describe drainage. Drainage rate may be assumed to follow a law of decreasing increments such that drainage rate becomes asymptotic to equilibrium infiltration rate. This can be expressed by the equation:

$$d = f_c - f_c e^{-kt} \qquad\qquad (2.31)$$

where d is the drainage rate (mm/h),

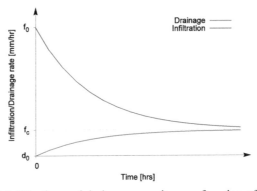

Figure 2-2: Infiltration and drainage capacity as a function of time

If infiltration is construed as input to soil water storages S and drainage as output, the continuity equation can be written

$$\frac{\Delta S}{\Delta t} = f - d \qquad (2.32)$$

Substitution of equations (2.30) and (2.31) in equation (2.32) yields

$$\frac{\Delta S}{\Delta t} = f_o e^{-kt} \qquad (2.33)$$

A storage mass curve is obtained by integrating equation (2.33)

$$S_t = \int_0^t \frac{\Delta S}{\Delta t} = \frac{f_0}{k}(1 - e^{-kt}) \qquad (2.34)$$

where S_t, is the accumulated soil water (mm) at time t (h).

Putting t equal to infinity in equation (2.34) yields the maximum soil water storage

$$S_{max} = \frac{f_0}{k} \qquad (2.35)$$

It may be noted that for each soil water state there is an associated potential rate of infiltration and a corresponding rate of drainage. Should the rate of precipitation, p, be lower than the potential rate of infiltration, f will be equal to p. The resulting change in soil water storage can be calculated from Eq. (2.32).

45

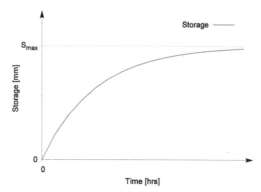

Figure 2-3: Soil water content as a function of time.

Operation of Modified Horton Equation

For the operation of a rainfall-runoff simulation model a time interval must be selected. The same time interval should be used for simulation of infiltration. Generally speaking the smaller the interval the better. To start the simulation an initial soil water content must be specified, appropriate to antecedent conditions. The time corresponding to any specified water storage, S_t, can be determined by solving equation (2.34) for t:

$$t = \frac{1}{k} ln\left(\frac{1}{1 - S_t k / f_o}\right) \qquad (2.36)$$

where t is the time *(h)* corresponding to a specific soil water content.

Infiltration and Excess Rain

The actual depth of infiltration accumulated during one time interval depends on the magnitude of f_t and $f_{t+\Delta t}$ relative to P_t, where f_t and $f_{t+\Delta t}$ are determined from equation (2.30). There are three possible case in relation to the rate of rainfall P and potential infiltration rate f_t.

Case 1: Rainfall rate, P_1 higher than or equal to potential infiltration rate, f_t. when the potential depth of infiltration is equal to the actual depth of infiltration the results obtained from the modified formula will be the same as would be given by Horton's original formula. The depth of infiltration is found by integrating equation (2.30) between t and $t+\Delta t$:

$$\bar{f} = f_c \Delta t + \frac{f_0 - f_c}{k}\left(e^{-kt} - e^{-k(t+\Delta t)}\right) \tag{2.37}$$

where $\bar{f}$ is the depth of infiltration (mm) infiltrated between time t and $t+\Delta t$.

The depth of excess rain for the same interval is then

$$\bar{e} = P_1 \Delta t - \bar{f} \tag{2.38}$$

where $\bar{e}$ e is the depth of excess rain (mm) and P_1 is the rainfall rate (mm/h) during the time interval.

Case 2: Rainfall rate, P_2 lower than f_t but higher than $f_{t+\Delta t}$. The actual depth of infiltration will be less than the potential infiltration depth. The time, t_p at which the potential infiltration rate becomes equal to the rate of rainfall, P, can be calculated by solving Eq. (2.30) for t with f set equal to P_2:

$$t = \frac{1}{k}\ln\left(\frac{f_0 - f_c}{P_2 - f_c}\right) \tag{2.39}$$

where t_p is the time (h) at which the infiltration rate f equals the precipitation rate P_2 (mm/h). The depth of infiltration during the time interval is then

$$\bar{f} = P_2\left(t_p - t\right) + f_c\left(t + \Delta t - t_p\right) + \frac{f_0 - f_c}{k}\left(e^{-kt_p} - e^{-k(t+\Delta t)}\right) \tag{2.40}$$

The depth of excess rain is then given by equation (2.38) with P_1 replaced by P_2.

Case 3: Rainfall rate, P_3 equal to or lower than potential infiltration rate $f_{t+\Delta t}$. In this case the depth of infiltration is given as;

$$\bar{f} = P_3 \Delta t \tag{2.41}$$

$$\bar{e} = 0 \tag{2.42}$$

Drainage

The depth of drainage can be found by integrating equation (2.31) over the given interval

$$\bar{d} = f_c \Delta t - \frac{f_c}{k}\left(e^{-kt} - e^{-k(t+\Delta t)}\right)$$ (2.43)

where $\bar{d}$ is the depth of drainage (mm).

Soil Water Storage

At the end of the interval of integration the change of soil water is given by

$$\Delta S = \bar{f} - \bar{d}$$ (2.44)

where ΔS is the change in soil water (mm) during the interval of integration.

The soil water storage at the beginning of the next time step S_{t+1} is obtained by adding ΔS to the previous soil water content.

$$S_{t+1} = S_t + \Delta S$$ (2.45)

2.9 Conclusion

This chapter provides the theoretical background required for understanding, development and application of urban flood modelling. It reviews the general forms, underlying assumptions, simplifications and applicability of the governing equations used to describe gradually varying free surface flows. It describes different types of models used for modelling shallow water flows and the numerical schemes to solve the shallow water equations. It also discusses the issues concerning modelling of urban floods and urban flood models, the use of topographical data for urban flood modelling and incorporation of infiltration modelling in urban flood models. The subsequent chapters present development and applications of methodologies for modelling of urban floods at different topographical resolutions.

3   Development of the Modelling Systems

3.1 Introduction

Urban flood modelling is undertaken using a hydrological model to determine the runoff for a rainfall event (or time series) over the urban surface, and a hydraulic model to simulate the flow through the urban drainage network and aboveground.

Numerical models of overland flow have been applied to a number of practical problems of interest in Engineering, including overland hydrology, open channel management and surface irrigation(Garcia-Navarro and Brufau, 2006). These types of numerical models are particularly interesting for the simulation of flood waves and their interaction with existing structures in urban flood modelling.

The one dimensional hydrodynamic modelling approach has been widely used in modelling flood flows due to its computational efficiency, ease of parameterization and easy representation of hydraulic structures in dealing with flows in large and complex networks both of conduits underground and of channels on the surface. However, one-dimensional (1D) models neglect some important aspects of the spatial variability of floodplain hydraulics and are too simplistic in their treatment of floodplain flows (Kuiry et al., 2010) thus the 1D assumption for modelling two-dimensional (2D) surface flow may be insufficient in many urban areas where the flow paths on the surface are often complicated to define because of crowded buildings, houses and roads (Mark et al., 2004).

Since the full 2D models are computationally more demanding than 2D models without convective momentum terms (i.e., noninertia models) many researchers and practitioners favour the use of Non-convective acceleration wave models over the full 2D models.

This chapter describes a newly developed 2D modelling system which employs an Alternating Direction Implicit (ADI) numerical procedure in combination with an iteration procedure, development of a coarse grid 2D modelling system which uses information derived from a fine grid resolution for the purpose of improving flood forecasts in geometrically complex urban environment, and the coupling of the developed 2D modelling system with a 1D sewer network modelling system (SWMM5) to simulate the complex nature of the interaction between surcharged sewer and flows associated with urban flooding.

3.2 Non-convective wave 2D overland flow model

3.2.1 Overland flow model

The Non-convective wave 2D modelling system represents the urban topography by the ground elevations at the centers and boundaries of cells on a rectangular Cartesian grid and determines the water levels at the cell centers and the discharges (velocities) at the cell boundaries. The alternating direction implicit finite difference procedure is used to solve the governing equations. The schematization of the topography coupled with the way the governing equations are solved allows a good representation of small-scale topographical elements in the urban environment including defined flow paths such as road networks and channels.

The system of 2D shallow water equations is obtained by integrating the Navier Stokes equations over depth and replacing the bed stress by a velocity squared resistance term in the two orthogonal directions. The continuity equation for the 2D flood plain flows is formulated as in Eq.(3.1):

$$\frac{\partial h}{\partial t} + \frac{\partial (hu)}{\partial x} + \frac{\partial (hv)}{\partial y} = 0 \tag{3.1}$$

Neglecting eddy losses, Coriolis force, variations in atmospheric pressure, wind shear effect and lateral inflow, the momentum equation can be written as in Eqs. (3.2) and (3.3).

In the x-direction:

$$\frac{\partial (hu)}{\partial t} + \frac{\partial (hu^2)}{\partial x} + \frac{\partial (huv)}{\partial y} + gh\frac{\partial H}{\partial x} + gC_f u\sqrt{u^2 + v^2} = 0 \tag{3.2}$$

In the y-direction

$$\frac{\partial (hv)}{\partial t} + \frac{\partial (huv)}{\partial x} + \frac{\partial (hv^2)}{\partial y} + gh\frac{\partial H}{\partial y} + gC_f v\sqrt{u^2 + v^2} = 0 \tag{3.3}$$

in which h is the water depth; H is the water level, u and v are the velocities in the directions of the two orthogonal axes (the x and y directions); g is the acceleration due to gravity, and the coefficient C_f appearing in the friction terms is normally expressed in terms of the Manning n or Chézy roughness factor (Garcia-Navarro and Brufau, 2006).

The terms in Eqs. (3.2) and (3.3) represent the various forces which control the propagation of flood waves as indicated in section 2.2.4 in the literature review. Various simplified flood flow models can be constructed, depending on which forces in the momentum equations are assumed negligible in comparison with the remaining forces. Hunter et al (2007) reviewed the application of simplified spatially-distributed models for predicting flood inundation. It was noted that whilst the neglect of the inertia terms leads to local inaccuracies, reduced complexity 2D schemes have been tested successfully against analytical solutions, results from physical or alternative numerical models and measurements from actual flood events. Two-dimensional flow over inundated urban flood plain is assumed to be a slow, shallow phenomenon and therefore the convective acceleration terms can be assumed to be sufficiently small compared to the other terms so that they can be ignored.

Expressing the velocities in terms of the discharges and using Chézy roughness factor, the simplified momentum equations can be written as:

In the x-direction

$$\frac{\partial}{\partial t}\left(\frac{Q}{Z_Q}\right)+\Delta Yg\frac{\partial h}{\partial x}+g\frac{Q}{C^2 Z_Q^2}\left(\left(\frac{1}{\Delta Y}\frac{Q}{Z_Q}\right)^2+\left(\frac{1}{\Delta X}\frac{R}{Z_R}\right)^2\right)^{0.5}=0 \qquad (3.4)$$

In the y-direction

$$\frac{\partial}{\partial t}\left(\frac{R}{Z_R}\right)+\Delta X\,g\frac{\partial h}{\partial y}+g\frac{R}{C^2 Z_R^2}\left(\left(\frac{1}{\Delta Y}\frac{Q}{Z_Q}\right)^2+\left(\frac{1}{\Delta X}\frac{R}{Z_R}\right)^2\right)^{0.5}=0 \qquad (3.5)$$

in which h is the water level; Q and R are the discharges in the directions of the two orthogonal axes (the x and y directions) ; ΔX and ΔY are the grid spacings in the X and Y directions; Z_Q and Z_R are the water depths at the cell boundaries, g is the acceleration due to gravity, and C is the Chézy friction factor.

3.2.2   Numerical solution, space and time discretization

The PDEs of the governing equations are transformed to difference equations on a regular Cartesian grid as shown in Figure 3-1 and a finite difference method is implemented for the numerical solution. A two point forward spatial and temporal difference scheme is adopted based on a uniform time step $\Delta t = t^{n+1} - t^n$, where n is time step counter.

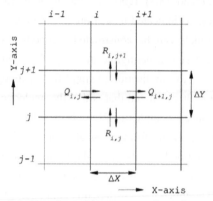

Figure 3-1: Regular grid representation of topography

Governing Equations in discretized form

The discretized form of the continuity and momentum equations is given in Eq. (3.6) to (3.9).

The continuity equation at the half time step:

$$0 = h_{i,j}^{n+\frac{1}{2}} - h_{i,j}^{n-\frac{1}{2}}$$

$$+ \frac{\Delta t}{\Delta X \Delta Y} \left[\theta \left(Q_{i+1,j} - Q_{i,j} \right)^{n+\frac{1}{2}} + (1-\theta) \left(Q_{i+1,j} - Q_{i,j} \right)^{n-\frac{1}{2}} \right] \qquad (3.6)$$

$$+ \frac{\Delta t}{\Delta X \Delta Y} \left[\left(R_{i,j+1} - R_{i,j} \right)^{n} \right]$$

The continuity equation at the full time step:

$$0 = h_{i,j}^{n+1} - h_{i,j}^{n}$$

$$+ \frac{\Delta t}{\Delta X \Delta Y} \left[\theta \left(R_{i,j+1} - R_{i,j} \right)^{n+1} + (1-\theta) \left(R_{i,j+1} - R_{i,j} \right)^{n} \right] \qquad (3.7)$$

$$+ \frac{\Delta t}{\Delta X \Delta Y} \left[\left(Q_{i+1,j} - Q_{i,j} \right)^{n+\frac{1}{2}} \right]$$

The momentum equation in the x-direction:

$$0 = \left(\frac{Q}{Z_Q}\right)^{n+\frac{1}{2}}_{i+1,j} - \left(\frac{Q}{Z_Q}\right)^{n-\frac{1}{2}}_{i+1,j}$$

$$+ g\frac{\Delta t \Delta Y}{\Delta X}\left[\theta\left(h_{i+1,j} - h_{i,j}\right)^{n+\frac{1}{2}} + (1-\theta)\left(h_{i+1,j} - h_{i,j}\right)^{n-\frac{1}{2}}\right] \qquad (3.8)$$

$$+ g\frac{\Delta t}{\Delta Y C^2}\left[\theta S_{fx}^{n+\frac{1}{2}} + (1-\theta)S_{fx}^{n-\frac{1}{2}}\right]$$

in which

$$S_{fx}^{n+\frac{1}{2}} = \left(\frac{Q}{Z_Q^2}\right)^{n+\frac{1}{2}}_{i+1,j}\left[\left(\left(\frac{Q}{Z_Q}\right)^{n+\frac{1}{2}}_{i+1,j}\right)^2 + \left(\left(\frac{R}{Z_R}\right)^n_{ave}\right)^2\right]^{0.5}$$

$$S_{fx}^{n-\frac{1}{2}} = \left(\frac{Q}{Z_Q^2}\right)^{n-\frac{1}{2}}_{i+1,j}\left[\left(\left(\frac{Q}{Z_Q}\right)^{n-\frac{1}{2}}_{i+1,j}\right)^2 + \left(\left(\frac{R}{Z_R}\right)^n_{ave}\right)^2\right]^{0.5}$$

$$\left(\frac{R}{Z_R}\right)^n_{ave} = \frac{1}{4}\left[\left(\frac{R}{Z_R}\right)^n_{i,j} + \left(\frac{R}{Z_R}\right)^n_{i+1,j} + \left(\frac{R}{Z_R}\right)^n_{i,j+1} + \left(\frac{R}{Z_R}\right)^n_{i+1,j+1}\right]$$

The momentum equation in the y-direction:

$$0 = \left(\frac{R}{Z_R}\right)^{n+1}_{i,j+1} - \left(\frac{R}{Z_R}\right)^n_{i,j+1}$$

$$+ g\frac{\Delta t \Delta X}{\Delta Y}\left[\theta\left(h_{i,j+1} - h_{i,j}\right)^{n+1} + (1-\theta)\left(h_{i,j+1} - h_{i,j}\right)^n\right] \qquad (3.9)$$

$$+ g\frac{\Delta t}{\Delta X C^2}\left[\theta S_{fy}^{n+1} + (1-\theta)S_{fy}^n\right]$$

in which

$$S_{fy}^{n+1} = \left(\frac{R}{Z_R^2}\right)^{n+1}_{i,j+1}\left[\left(\left(\frac{R}{Z_R}\right)^{n+1}_{i,j+1}\right)^2 + \left(\left(\frac{Q}{Z_Q}\right)^{n+\frac{1}{2}}_{ave}\right)^2\right]^{0.5}$$

$$S_{fy}^{n} = \left(\frac{R}{Z_R^2}\right)_{i,j+1}^{n} \left[\left(\left(\frac{R}{Z_R}\right)_{i,j+1}^{n}\right)^2 + \left(\left(\frac{Q}{Z_Q}\right)_{ave}^{n-\frac{1}{2}}\right)^2\right]^{0.5}$$

$$\left(\frac{Q}{Z_Q}\right)_{ave}^{n+\frac{1}{2}} = \frac{1}{4}\left[\left(\frac{Q}{Z_Q}\right)_{i,j}^{n+\frac{1}{2}} + \left(\frac{Q}{Z_Q}\right)_{i+1,j}^{n+\frac{1}{2}} + \left(\frac{Q}{Z_Q}\right)_{i,j+1}^{n+\frac{1}{2}} + \left(\frac{Q}{Z_Q}\right)_{i+1,j+1}^{n+\frac{1}{2}}\right]$$

Here Δt is the time step, θ is time weighting coefficient, S_{fx} and S_{fy} are the friction slopes in x and y directions respectively and n is the time step counter.

Solving the governing equations

The ADI procedure is used to solve the governing equations for flood flows. The main characteristic of ADI algorithm is the splitting of the calculation into two series of 'one - dimensional' calculations which are mutually orthogonal (Kutija, 1996). According to (Peaceman and Rachford, 1955) the method provides greater superiority to the explicit finite difference method due to the high computational efficiency which requires less computing time because it involves a tridiagonal matrix and uses larger time steps due to its unconditional stability (provided the finite difference scheme is properly formulated). The explicit difference equation is simple to solve but it requires an uneconomically large number of time steps of undefined size and the full implicit difference equations do not limit the time step but they require a complex matrix inversion that is itself time consuming at each time step for the solution of large set of simultaneous equation. In the ADI algorithm the solution procedure is split in such a way that in one direction the conservation of mass and conservation of the momentum in that direction are solved, and, after that, in the other direction, the conservation of mass is again solved but now with the conservation of momentum for this direction. For this purpose, the time step is divided into two parts such that Q is defined at the half time step and R at the whole time step.

Therefore the equations are solved sequentially in the x and y directions in two half time steps, $t^{n+1/2}$ and t^{n+1} respectively. A schematic representation of the ADI algorithm is shown in Figure 3-2.

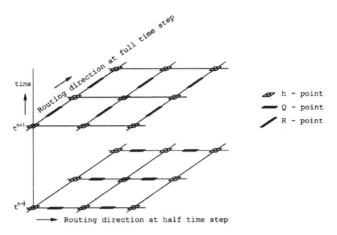

Figure 3-2: Schematic representation of alternating direction flood routing during each half time step

Multiplying the momentum equation in the x-direction by the depth $Z_{Qi+1,j}^{n+1/2}$ at the cell boundary where $Q_{i+1,j}$ is defined, the residual can be written as;

$$ResQ_{i+1,j}^{n+\frac{1}{2}} = Q_{i+1,j}^{n+\frac{1}{2}} - Z_{Qi+1,j}^{n+\frac{1}{2}} * \left(\frac{Q}{Z_Q}\right)_{i+1,j}^{n-\frac{1}{2}}$$

$$+ Z_{Qi+1,j}^{n+\frac{1}{2}} g \frac{\Delta t \Delta Y}{\Delta X}\left[\theta\left(h_{i+1,j} - h_{i,j}\right)^{n+\frac{1}{2}} +(1-\theta)\left(h_{i+1,j} - h_{i,j}\right)^{n-\frac{1}{2}}\right] \quad (3.10)$$

$$+ Z_{Qi+1,j}^{n+\frac{1}{2}} g \frac{\Delta t}{\Delta YC^2}\left[\theta S_{fx}^{n+\frac{1}{2}} +(1-\theta)S_{fx}^{n-\frac{1}{2}}\right]$$

which can be written as a function of increments in the three unknown variables $h_{i,j}^{n+\frac{1}{2}}$, $h_{i+1,j}^{n+\frac{1}{2}}$, and $Q_{i+1,j}^{n+\frac{1}{2}}$ as

$$ResQ_{i+1,j}^{*n+\frac{1}{2}} = \frac{\partial}{\partial h_{i,j}^{n+\frac{1}{2}}}\left(ResQ_{i+1,j}^{n+\frac{1}{2}}\right)\Delta h_{i,j}^{n+\frac{1}{2}} + \frac{\partial\left(ResQ_{i+1,j}^{n+\frac{1}{2}}\right)}{\partial Q_{i+1,j}^{n+\frac{1}{2}}}\Delta Q_{i+1,j}^{n+\frac{1}{2}}$$

$$+ \frac{\partial}{\partial h_{i+1,j}^{n+\frac{1}{2}}}\left(ResQ_{i+1,j}^{n+\frac{1}{2}}\right)\Delta h_{i+1,j}^{n+\frac{1}{2}} \quad (3.11)$$

Similarly the residual of the continuity equation at the half time step is,

$$Resh_{i,j}^{n+\frac{1}{2}} = h_{i,j}^{n+\frac{1}{2}} - h_{i,j}^{n-\frac{1}{2}}$$

$$+\frac{\Delta t}{\Delta X \Delta Y}\left[\theta\left(Q_{i+1,j} - Q_{i,j}\right)^{n+\frac{1}{2}} + \left(1-\theta\right)\left(Q_{i+1,j} - Q_{i,j}\right)^{n-\frac{1}{2}}\right] \quad (3.12)$$

$$+\frac{\Delta t}{\Delta X \Delta Y}\left[\left(R_{i+1,j} - R_{i,j}\right)^{n}\right]$$

$$Resh^{*n+\frac{1}{2}}_{i,j} = \frac{\partial}{\partial h_{i,j}^{n+\frac{1}{2}}}\left(Resh_{i,j}^{n+\frac{1}{2}}\right)\Delta h_{i,j}^{n+\frac{1}{2}}$$

$$\qquad\qquad (3.13)$$

$$+\frac{\partial}{\partial Q_{i,j}^{n+\frac{1}{2}}}\left(Resh_{i,j}^{n+\frac{1}{2}}\right)\Delta Q_{i,j}^{n+\frac{1}{2}} + \frac{\partial}{\partial Q_{i+1,j}^{n+\frac{1}{2}}}\left(Resh_{i,j}^{n+\frac{1}{2}}\right)\Delta Q_{i+1,j}^{n+\frac{1}{2}}$$

Solving for the increments in the discharges $\Delta Q_{i,j}^{n+\frac{1}{2}}$ and $\Delta Q_{i+1,j}^{n+\frac{1}{2}}$ from their respective residue equations and substituting in Eq.(3.13), the residual of the continuity equation can be written as a function of only the unknown water levels $h_{i-1,j}^{n+\frac{1}{2}}$, $h_{i,j}^{n+\frac{1}{2}}$ and $h_{i+1,j}^{n+\frac{1}{2}}$ as in Eq. (3.14).

$$Resh^{*n+\frac{1}{2}}_{i,j} = \frac{\partial}{\partial h_{i-1,j}^{n+\frac{1}{2}}}\left(Resh_{i,j}^{n+\frac{1}{2}}\right)\Delta h_{i-1,j}^{n+\frac{1}{2}}$$

$$\qquad\qquad (3.14)$$

$$+\frac{\partial}{\partial h_{i,j}^{n+\frac{1}{2}}}\left(Resh_{i,j}^{n+\frac{1}{2}}\right)\Delta h_{i,j}^{n+\frac{1}{2}} + \frac{\partial}{\partial h_{i+1,j}^{n+\frac{1}{2}}}\left(Resh_{i,j}^{n+\frac{1}{2}}\right)\Delta h_{i+1,j}^{n+\frac{1}{2}}$$

where the Δh's are the differences in the water level in the corresponding cells. At any time step the equations for the residual of the continuity equation (Eq. (3.14)) lead to a system of algebraic equations with a tridiagonal matrix of coefficients which can be solved by forward sweep and a backward substitution. The algorithm is simply based on Gaussian elimination.

The new water level in $cell(i,j)$ is then computed as $h_{i,j}^{n+\frac{1}{2}} = h^{*n+\frac{1}{2}}_{i,j} - \Delta h_{i,j}^{n+\frac{1}{2}}$ where $h^{*n+\frac{1}{2}}_{i,j}$ is the water level of $cell(i,j)$ from the previous iteration. Iterations are carried out until the change in water level differences between successive iterations is less than a preset convergence tolerance or the maximum predefined number of iteration is reached. If the iteration does not converge and the number of predefined iteration is reached, the iteration is repeated with a new time step which is half the previous time step.

The change in discharge is calculated from the residual in momentum equation (Eq. (3.11)) as shown in Eq. (3.15).

$$\Delta Q_{i+1,j}^{n+\frac{1}{2}} = \frac{\left(ResQ_{i+1,j}^{n+\frac{1}{2}} - \frac{\partial ResQ_{i+1,j}^{n+\frac{1}{2}}}{\partial h_{i,j}^{n+1}} \Delta h_{i,j}^{n+\frac{1}{2}} - \frac{\partial ResQ_{i+1,j}^{n+\frac{1}{2}}}{\partial h_{i+1,j}^{n+1}} \Delta h_{i+1,j}^{n+\frac{1}{2}} \right)}{\dfrac{\partial ResQ_{i+1,j}^{n+\frac{1}{2}}}{\partial Q_{i+1,j}^{n+\frac{1}{2}}}} \tag{3.15}$$

The updated discharge to $cell(i,j)$ is then computed as $Q_{i,j}^{n+\frac{1}{2}} = Q_{i,j}^{*n+\frac{1}{2}} - \Delta Q_{i,j}^{n+\frac{1}{2}}$, where $Q_{i,j}^{*n+\frac{1}{2}}$ is the discharge to $cell(i,j)$ from previous iteration. Similarly, for the next half time step, the equations are solved for the y-direction. Figure 3-5 shows the solution algorithm of the developed Non-convective wave 2D flood modelling system in the form of pseudo-code.

Time Stepping

The modelling system has the ability to halve or double the time step depending on the convergence tolerance of the solution in order to have adaptable time steps for efficient computation. The time stepping is carried out in such a way that it doubles if the model runs for some defined amount of time step without changing the time step or it halves if the convergence criteria is not satisfied within the maximum number of iterations. Iterations are carried out until the change in water level differences between successive iterations is less than a preset convergence tolerance or the maximum predefined number of iteration is reached. If the iteration does not converge and the number of predefined iteration is reached, the iteration is repeated with a new time step which is half the previous time step. The computations of the discharge and at the cell borders and the water levels values for the next time step make use of their corresponding vales at the current and previous time steps. When the computational time step changes (either doubles or halves), the location of the previous time step changes and therefore the values of the discharges and water levels are not readily available. These values need to be estimated to proceed with the computation for the next time step. In the current modelling system, quadratic interpolation is used to estimate the discharges and water levels at previous time steps when there is a time step change. The values need to be estimated using quadratic interpolation are shown in Figure 3-3.

The flow depth at the boundary of adjacent cells and the derivatives of the flow depth with respect to the water levels of the adjacent cells are evaluated in relation to the ground surface elevation at the boundary and the water levels of the cells. Figure 3-4 shows the representation of the water levels of cells, the flow depth at the boundary and cell elevations. There are four possible cases for the relative relationship of the water level of adjacent cells and the ground elevation at the boundary of the adjacent cells. Table 3-1 shows the equation used to compute the flow depth for each case. In Figure 3-4, h is water level, D is the ground surface elevation at the boundary of $cell(i, j)$ and $cell(i+1, j)$, which is equal to the maximum ground surface elevation of the two adjacent cells, fd is the flow depth at the boundary of the two cells. The flow depth at the boundary of $cell(i, j)$ and $cell(i+1, j)$ is calculated according to the water levels of the cells in relation to the ground surface elevation at the boundary (D).

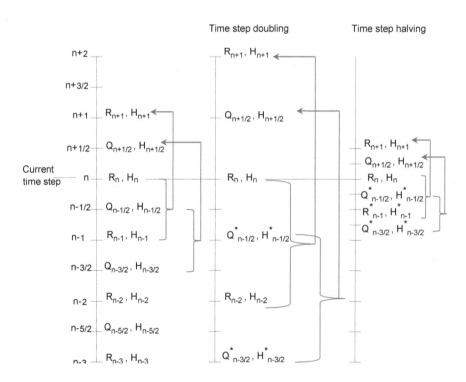

Figure 3-3: Computational progress during time step doubling and halving - the values Q*, R* and H* are estimated by quadratic interpolation

Wetting and drying

The water depth of a grid cell is calculated as the average depth over the whole cell. When the cell first receives water, the wetting front edge usually lies within the cell. In most cases, only part of the cell will be wetted at that time step. According to Yu and Lane (2006) if this problem is not dealt with, then water will diffuse too quickly across the floodplain. When the flow volume leaving a cell is greater than that entering the cell, the cell is drying and there is the possibility that the water depth may be reduced to zero or a negative value. This presents two separate problems in the model. Clearly, negative depths are impossible. However, more seriously, it is vital that partially wet cells are maintained accurately during drying to avoid the creation of isolated artificially wet patches. The model used here follows a modification of the principle used by Yu and Lane (2006). The wetting process is controlled by a wetting parameter. When the cell is wetting, the water should not be allowed to flow out of the cell until the wetting front has crossed the cell. Following Yu and Lane (2006), each cell has a property called '%wet'. When the cell is first wetted,

$$\%wet = min\left(1, \frac{\sum(v\Delta t)}{\Delta X}\right) \qquad (3.16)$$

where v is the velocity calculated from the discharge crossing the cell boundary divided by the cell width and the cell flow depth, ΔX is the cell width and Δt is the current time step. Water is not allowed to flow out of the cell until the wetting parameter reaches unity (i.e. the cell is fully wet). The wetting parameter is updated in each time step to describe the water traveling across a cell. As (Yu and Lane, 2006 (a)) indicated this parameter is a necessary, yet not sufficient, condition for water to flow out of the cell. A minimum water depth is set. Before this depth is reached, no outflow is allowed.

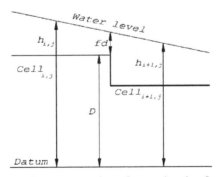

Figure 3-4: Schematic representation of water levels of cells, flow depth
at the boundary and cell elevations

Table 3-1: Computation of flow depths at the boundary of adjacent cells

Case	Description	Flow depth
1	both $h_{i,j}$ and $h_{i+1,j}$ are greater than D	$fd = 0.5*\left(h_{i,j} + h_{i+1,j}\right) - D$
2	$h_{i,j} > D$ and $h_{i+1,j} < D$	$fd = 0.5*\left(h_{i,j} - D\right)$
3	$h_{i,j} < D$ and $h_{i+1,j} > D$	$fd = 0.5*\left(h_{i+1,j} - D\right)$
4	both $h_{i,j}$ and $h_{i+1,j}$ are less or equal to D	$fd = 0$

During drying to determine when to remove the cell from the calculations, a minimum depth (dmin) is set. When the water depth is below or equal to the minimum depth, no outflow is allowed from the cell. If the net outflow is such that the depth in the cell would be reduced below the minimum depth, then the outflow is adjusted to be the volume of water available in the cell per the time step

Stability of the 2D model

The stability analysis for a numerical solution obtained by any finite-difference method is usually done by means of a linearised analysis in Fourier series expansions (Kutija, 1996). A stability analysis of a similar discretisation used for the 2D model is given in (Casulli, 1990)with the inertia term included and he showed that the stability of the method is believed to be guaranteed for practical purposes, if the inertia terms are dropped. Kutjia (1996) argued such analysis is highly localized and does not consider the application of the boundary conditions and, although stability proved in this way is essential for the performance of any scheme, it is by no means sufficient, and this is especially so in the case of implicit schemes. Several numerical experiments on different hypothetical case studies show that the Non-convective wave 2D model provides stable results even for Courant numbers much greater than 1. Although the numerical scheme used is believed to be stable for practical purposes, the time step is limited for accuracy purposes. The model uses adaptive time step for increasing model efficiency. It has the ability to halve or double the time step; halving in order to meet the convergence criterion, and doubling after a certain number of time steps without halving.

```
read data from input files and initialize
loop on all time steps
begin
        if doubling time step
                double time step
        else
                continue

        if time step changes
                compute previous values by quadratic interpolation
        else
                continue

        start iteration
                loop on all x-direction lines
                begin
                        values of the boundaries
                        calculate coefficients
                        solution of this line by the double-sweep
                end

                if convergence criteria satisfied
                        continue
                else
                        re-start iteration with new time step equal to half the previous

                loop on all y-direction lines
                begin
                        values of the boundaries
                        calculate coefficients
                        solution of this line by the double-sweep
                end

                if convergence criteria satisfied
                        continue
                else
                        re-start iteration with new time step equal to half the previous

end
```

Figure 3-5: Solution algorithm pseudo-code

3.3   Generalization from fine to coarse grid

There are several modelling techniques for urban inundation modelling. The choice depends on the type of representation required. If only a rough estimation of the flood characteristics such as water levels and velocities at a coarse grid is needed, then a simple coarse grid model which represents the overall increase in flow resistance due to the presence of buildings and urban structures with a high friction coefficient may be good enough. However this approach does not provide a local description of the flow even with fine grid resolution because the local effects of the urban features are not present.

A step further with this approach is the introduction of the concepts of Urban Porosity, which define the fraction of the total surface that the area subject to flooding in terms porosity. Researchers have proposed modified shallow water models with porosity to account for the presence of buildings, structures etc. that restrict the area available to water flow (Guinot and Soares-Frazão, 2006; Hervouet *et al.*, 2000). This approach affects the mass conservation equation.

These techniques are best suited for modelling large areas where it is impractical to use high resolution topographic data, and are successful in simulating flood extent and the flood duration, provided friction coefficients are reasonably estimated.

Yu and Lane (2006 (a)) developed a method for sub-grid-scale topographic representation with an explicit treatment of the effects of structural elements on the floodplain both in terms of blockage and flux. They represent cell flux effects using a porosity-type treatment with a correction for the water depth since the flux is a non-linear function of the water depth.

Without a sub-grid-scale topographic representation, the elevation E of the cell can be calculated simply as the average of the sub-grid cells. The relationship between water surface elevation and the volume of water that can be stored within the cell without considering the sub-grid topography is a simple linear one. The relationship between water surface elevation and the flow area is also linear.

These linear relationships should be modified to represent the actual volume storage effect and the flow area effect. To address the problem of capturing small-scale urban features in a coarse resolution 2D model with the aim of improving flood forecasts in geometrically complex urban environment, the continuity and momentum equations which describe the 2D model as given in Eqs. (3.1), (3.4) and (3.5) are rewritten in such a way that the volume and area are expressed as a non-simple function of water depth (instead of a fixed plane area of the coarse grid). We reinterpret the original equations defined for flow at a point in terms of a finite cell. The water depth and velocities can be described by Eqs (3.17) to Eqs (3.19).

$$\frac{\partial h}{\partial t} + \Delta x \frac{\partial \left(u A_Q \right)}{\partial x} + \Delta y \frac{\partial \left(v A_R \right)}{\partial y} = q \qquad (3.17)$$

$$\frac{\partial}{\partial t}\left(\frac{Q}{A_Q}\right)+g\frac{\partial h}{\partial x}+g\frac{1}{C^2 Z_Q}\frac{Q}{A_Q}\left[\left(\frac{Q}{A_Q}\right)^2+\left(\frac{R}{A_R}\right)^2\right]^{0.5}=0 \quad x-direction\ (3.18)$$

$$\frac{\partial}{\partial t}\left(\frac{R}{A_R}\right)+g\frac{\partial h}{\partial y}+g\frac{1}{C^2 Z_R}\frac{R}{A_R}\left[\left(\frac{Q}{A_Q}\right)^2+\left(\frac{R}{A_R}\right)^2\right]^{0.5}=0 \quad y-direction\ (3.19)$$

in which, Q and R are the cell discharges in the directions of the two orthogonal axes ; $u=\left(Q/A_Q\right)$ and $v=\left(R/A_R\right)$ are flow velocities at the cell boundaries (Q and R); A_Q and A_R are flow areas at the cell boundaries (Q and R); Δx and Δy are the grid spacing in the x and y directions; q is source or sink and t is flow time, g is the acceleration due to gravity, Z_Q and Z_R are water depths at cell boundaries and C is the Chézy friction factor.

As explained by Yu and Lane (2006 (b)), the relationship between water surface elevation and volume of water that can be stored with in the cell is a simple linear one when the average elevation is considered for the coarse grid. Consider the example of a coarse grid cell and its fine grid cells (sub-grid cells) in the configuration shown in Figure 3-6. The coarse cell is composed of nine sub-grid cells with bed elevations E1, E2, and up to E9 (Figure 3-6 (a)). In the coarse grid representation, the elevation E of the cell can be calculated simply as the average of the nine sub-grid cells. The relationship between water surface elevation and the volume of water that can be stored within the cell without considering the fine grid cells is a simple linear one, taking the form;

$$V_{ij}\left(H_{ij}\right)=w^2\left(H_{ij}-\frac{1}{N_{ij}}\sum_{k=1}^{N_{ij}^k}E_k\right) \quad (3.20)$$

where $V_{ij}\left(H_{ij}\right)$ is the volume of water that can be stored in the coarse grid cell ij for water surface elevation H_{ij}, w is the coarse grid resolution, N_{ij} is the number of fine grid cells contained in the course grid (sub-grid cells) in grid cell ij and E_k is the bed elevation of each sub-grid cell.

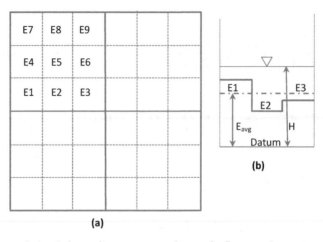

Figure 3-6: Schematic representation of fine and coarse grid topography, E1, E2, ..., E9 are the elevation of the fine grid cells and Eavg is the average of the fine grid cell elevations. H is the coarse grid cell water surface elevation

Calculating the volume stored using equation Eq. (3.20) could underestimate (if the water level is between the minimum and maximum fine grid cell elevations) or overestimate (if the water level is above the maximum fine grid cell elevation) the volume. The flow areas in all directions will be affected the same way. However, rewriting the continuity and momentum equations in the form of Eqs (3.17) to (3.19) enables us to take the actual volume stored as in Eq(3.21) (with a uniform water level across the cell) and flow areas, which can be computed from the fine grid resolution and stored in look-up tables for the volume - depth and area of flow – depth relationships. These relationships can be computed for any coarse grid resolution as in Eqs (3.21) and (3.22) respectively.

$$V_{ij}\left(H_{ij}\right) = w^2\left(H_{ij} - \frac{1}{N_{ij}}\sum_{k=1}^{N_{ij}^k} E_k\right) \tag{3.21}$$

where N_{ij}^k is the number of wet fine grid cells contained in the coarse grid (sub-grid cells) in grid cell ^{ij}.

$$A_{Qij}\left(H_{ij}\right) = \frac{w}{Nx_{ij}}\sum_{k=1}^{Nx_{ij}^k}\left(H_{ij} - E_k\right) \tag{3.22}$$

where $A_{Qij}\left(H_{ij}\right)$ is the area of flow in the x-direction, Nx_{ij} is the number of border fine grid cells in the x-direction, and Nx_{ij}^{k} the number of wet border cells in the x-direction. The area of flow in the y-direction is also computed in similar fashion.

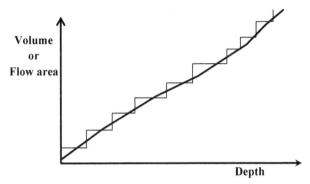

Figure 3-7: Depth-volume or depth-flow area relationship

The volume-depth and area of flow-depth relationships are evaluated from the finer grid available for each coarse grid cell and stored in a look-up table. The method retains computational efficiency by establishing the relationships with a pre-processing step such that the values are evaluated using linear interpolation during the model solution.

3.4 Coupling the 2D Inundation Modelling System with the 1D-SWMM System

In most cases, models used for reproducing floods in urban areas have a sound physical basis in terms of the shallow-water equations, and their reliability increases with the amount of measurement data that can be used for their instantiation and calibration (or validation). A one-dimensional (1D) approach has been used as a standard industry practice for more than 30 years. Consequently there are still many researchers and practitioners who favour the 1D over other approaches. Some of the reasons for this are that these models are relatively easy to set up, calibrate and explain. To date, a large number of 1D models have been developed. These include both advanced commercial software products such as MOUSE (DHI Group, 2009a) and InfoWorks-CS and open source software like the Storm Water Management Modelling (SWMM) system developed by the U.S. Environmental Protection Agency (Rossman *et al.*, 2005); see (Hsu *et al.*, 2000). Most of these models use the water stage-volume curve to determine the flood depth caused by manhole surcharge and flooding. Although some of these models can replicate the interaction with surface channels it is generally the case that 1D models cannot properly simulate the complex phenomenon of the interaction between the flows in the sewers and above ground. Furthermore, 1D models neglect some important aspects and suffer from a number of drawbacks when applied to floodplain flows, including the inability to simulate lateral diffusion of the flood wave, the discretization of topography as cross-sections rather than as a surface and the subjectivity of cross-section location and orientation (Hunter *et al.*, 2007; Kuiry *et al.*, 2010). Also, in many cases the 1D assumption for modelling two-dimensional (2D) surface flow has been found inadequate as the flow paths along urban floodplains are difficult to define due to the complex geometry formed by buildings, houses and roads; see for example (Mark *et al.*, 2004).

More recently, fully integrated 1D-1D and 1D-2D approaches have been receiving greater attention (Chen *et al.*, 2012; Djordjević *et al.*, 2005; Leandro *et al.*, 2009; Leandro *et al.*, 2009). Several studies have been undertaken to evaluate potential and limitations of 1D and 1D-2D models (Mark *et al.*, 2004; Vojinovic and Tutulic, 2009). These studies suggest that 1D above ground models are economical, robust and preferred alternatives as long as the flood water remains in the street profile and as long as the overland flow paths can be identified. However, during heavy flooding the 1D approximation may be insufficient and the use of 2D models to describe the surface flow is preferred.

The development and use of coupled 1D-2D models for urban flood modelling has been driven by at least two objectives. The first is the need to assess more adequately the performance of storm or combined sewer systems and their associated surface flooding in order to predict the value of potential flood damages and to provide decision makers with information to design appropriate measures. The other objective is the development of 2D flood models that are capable of providing better predictions of the extent of flooding with respect to the depth-averaged Navier-Stokes equations. With these objectives in mind, current modelling of urban flood flows utilises a combination of one and two-dimensional hydraulic models. This approach enables elements which are essentially one-dimensional (e.g., drains, culverts, channels) to be modelled explicitly, while overland processes are modelled with a two-dimensional schematization. The results from hydrological rainfall-runoff simulation are used as inputs to the hydraulic model at specified manholes (junction nodes) of the drainage network. When the capacity of the pipe network is exceeded, excess flow spills from the manholes into the domain of a two-dimensional model.

The developed 2D model as described in section 4.1 is coupled with SWMM5 to simulate the complex nature of the interaction between surcharged sewer and flows associated with

urban flooding. The following is a description of the details of coupling this model with
the 1D sewer network model (SWMM5).

3.4.1   Sewer network model

In the present work, the Storm Water Management Model (SWMM5) developed by the
United States Environmental Protection Agency (USEPA) (Rossman *et al.*, 2005) is used
to simulate flows in a storm sewer system.

SWMM5 solves the conservation of mass and momentum equations (the Saint Venant
equations) that govern the unsteady flow of water through a drainage network of channels
and pipes by converting the equations into an explicit set of finite difference equations.

The Saint Venant equations and their solution method as implemented in SWMM5 are
described in Rossman (2006) The description in the following paragraphs are taken from
Rossman (2006).

The Saint Venant equations are expressed in the following form for flow along an
individual conduit.

$$\frac{\partial A}{\partial t} + \frac{\partial Q}{\partial x} = 0 \tag{3.23}$$

$$\frac{\partial Q}{\partial t} + \frac{\partial \left(Q^2/A\right)}{\partial x} + gA\frac{\partial H}{\partial x} + gAS_f + gAh_L = 0 \tag{3.24}$$

where x is distance along the conduit, t is time, A is cross-sectional area, Q is flow rate, H
is the hydraulic head of water in the conduit (elevation head plus any possible pressure
head), S_f is the friction slope (head loss per unit length), h_L is the local energy loss per unit
length of conduit, and g is the acceleration of gravity. For a known cross-sectional
geometry, the area A is a known function of flow depth y which in turn can be obtained
from the head H. Thus the dependent variables in these equations are flow rate Q and
head H, which are functions of distance x and time t.

When analysing a network of conduits, an additional continuity relationship is needed for
the junction nodes that connect two or more conduits together. In SWMM a continuous
water surface is assumed to exist between the water elevation at the node and in the

conduits that enter and leave the node (with the exception of free fall drops should they occur). The change in hydraulic head H at the node with respect to time can be expressed as:

$$\frac{\partial H}{\partial t} = \frac{\sum Q}{Astore + \sum As} \tag{3.25}$$

where $Astore$ is the surface area of the node itself, $\sum As$ is the surface area contributed by the conduits connected to the node, and $\sum Q$ is the net flow into the node (inflow – outflow) contributed by all conduits connected to the node as well as any externally imposed inflows.

Equations (3.23), (3.24), and (3.25) are solved in SWMM by converting them into an explicit set of finite difference formulas that compute the flow in each conduit and head at each node for time $t + \Delta t$ as functions of known values at time t. The equation solved for the flow in each conduit is:

$$Q_{t+\Delta t} = \frac{Q_t + \Delta Q_{gravity} + \Delta Q_{inertial}}{1 + \Delta Q_{friction} + \Delta Q_{losses}} \tag{3.26}$$

The individual ΔQ terms have been named for the type of force they represent and are given by the following expressions:

$$\Delta Q_{gravity} = g\bar{A}\left(H_1 - H_2\right)\Delta t / L$$

$$\Delta Q_{inertail} = 2\bar{V}\left(\bar{A} - A_1\right) + \bar{V}^2\left(A_2 - A_1\right)\Delta t / L$$

$$\Delta Q_{friction} = \frac{gn^2 \left|\bar{V}\right| \Delta t}{k^2 \bar{R}^{4/3}}$$

$$\Delta Q_{losses} = \frac{\sum_i K_i \left|V_i\right| \Delta t}{2L}$$

where:

$\bar{A}$ = average cross-sectional flow area in the conduit,

$\bar{R}$ = average hydraulic radius in the conduit,

V = average flow velocity in the conduit,

V_i = local flow velocity at location i along the conduit,

K_i = local loss coefficient at location i along the conduit,

H_1 = head at upstream node of conduit,

H_2 = head at downstream node of conduit,

A_1 = cross-sectional area at the upstream end of the conduit,

A_2 = cross-sectional area at the downstream end of the conduit.

The equation solved for the head at each node is:

$$H_{t+\Delta t} = H_t + \frac{\Delta Vol}{\left(Astore + As\right)_{t+\Delta t}} \tag{3.27}$$

where ΔVol is the net volume flowing through the node over the time step as given by:

$$\Delta Vol = 0.5\left[\left(\sum Q\right)_t + \left(\sum Q\right)_{t+\Delta t}\right]\Delta t$$

SWMM 5 solves equations (3.26) and (3.27) using a method of successive approximations with under relaxation. The procedure outlined below:

1. A first estimate of flow in each conduit at time $t+\Delta t$ is made by solving Eq. (4) using the heads, areas, and velocities found at the current time t. Then the same is done for heads by evaluating Eq. (5) using the flows just computed. These solutions are denoted as Q_{last} and H_{last}.

2. Eq. (4) is solved once again, using the heads, areas, and velocities that belong to the Q_{last} and H_{last} values just computed. A relaxation factor Ω is used to combine the new flow estimate Q_{new}, with the previous estimate Q_{last} according to the equation $Q_{new} = (1-\Omega)Q_{last} + \Omega Q_{new}$ to produce an updated value of Q_{new}.

3. Eq. (5) is solved once again for heads, using the flows Q_{new}. As with flow, this new solution for head, H_{new}, is weighted with H_{last} to produce an updated estimate for heads, $H_{new} = (1-\Omega)H_{last} + \Omega H_{new}$.

4. If H_{new} is close enough to H_{last} then the process stops with Q_{new} and H_{new} as the solution for time $t+\Delta t$. Otherwise, H_{last} and Q_{last} are replaced with H_{new} and Q_{new}, respectively, and the process returns to step 2.

In implementing this procedure, SWMM 5 uses a constant relaxation factor Ω of 0.5, a convergence tolerance of 0.005 feet (0.0015 meter) on nodal heads, and limits the number of trials to four.

Surcharge conditions
In SWMM surcharging occurs when all pipes entering a node are full or the water surface at the node lies between the crown of the highest entering pipe and the ground surface. Flooding is a special case of surcharge which takes place when the hydraulic grade line is

above the ground surface and water is lost from the sewer node to the aboveground system.

The amount of surcharged flow can be defined as $Q_s = Q_{in} + Q_r - Q_f$, where Q_{in} is the total inflow discharge from the upstream conduits, Q_r is the surface runoff coming in to the manhole; and Q_f is the design full capacity of the downstream conduits, defined as

$$Q_f = \frac{1}{n} A_f R_f^{2/3} S_f^{1/2}$$

Here n is the Manning's roughness of the conduit, A_f is the full cross section area of the conduit, R_f is the hydraulic radius for full conduit flow, and S_f is the friction slope of the conduit.

For the condition when the total inflow discharge ($Q_{in} + Q_r$) does not exceed the design outflow capacity (Q_f) as shown in Figure 3-8 (a) the actual discharge in the conduit ($Q_c = Q_{in}$) is drained through the downstream conduit without surcharge. When the conduit surcharges, as shown in Figure 3-8 (b) only a part of the inflow discharge equal to the pipe full capacity ($Q_c = Q_f$) can be drained by the downstream outflow conduit. The rest of the discharge ($Q_s = Q_{in} - Q_f$) which is in excess of the design capacity, surcharges through the manhole onto the ground surface (Hsu et al., 2000). If the volume due to this surcharge exceeds the manhole full volume capacity in a time step, the difference of these two volumes will be discharged as overflow to the surface.

Under surcharge condition the surface area contributed by any closed conduits would be zero and Eq. (3.25) would no longer be applicable. To accommodate this situation, SWMM uses an alternative nodal continuity condition, namely that the total rate of outflow from a surcharged node must equal the total rate of inflow, $\Sigma Q = 0$. By itself, this equation is insufficient to update nodal heads at the new time step since it only contains flows. In addition, because the flow and head updating equations for the system are not solved simultaneously, there is no guarantee that the condition will hold at the surcharged nodes after a flow solution has been reached.

To enforce the flow continuity condition, it can be expressed in the form of a perturbation equation:

$$\sum \left[Q + \frac{\partial Q}{\partial H} \Delta H \right] = 0$$

where ΔH is the adjustment to the node's head that must be made to achieve flow continuity. Solving for ΔH yields:

$$\Delta H = \frac{-\sum Q}{\sum \partial Q / \partial H} \qquad (3.28)$$

where from Eq. (3.26),

$$\frac{\partial Q}{\partial H} = \frac{-g\overline{A}\,\Delta t / L}{1 + \Delta Q_{friction} + \Delta Q_{losses}}$$

$\partial Q / \partial H$ has a negative sign in front of it because when evaluating $\sum Q$, flow directed out of a node is considered negative while flow into the node is positive.

Every time that Eq. (3.28) is applied to update the head at a surcharged node, Eq. (3.26) is re-evaluated to provide flow updates for the conduits that connect to the node. This process continues until some convergence criterion is met. These surcharge iterations are folded into its normal set of iterations outlined previously. That is, whenever heads need to be computed in the successive approximation scheme, Eq. (3.28) is used in place of Eq. (3.27) if a node is surcharged, and no under relaxation of the resulting head value is performed.

3.4.2   Model linkage

Two distinct models are combined for simulating the flow dynamics in sewer networks and on the aboveground surface.

The hydrological rainfall-runoff process and routing of flows in drainage pipes are performed using the 1D sewer network model EPA SWMM. When the capacity of the pipe network is exceeded, excess flow spills into the two-dimensional model domain from the manholes and is then routed using the Non-convective wave 2D overland flow model. Both models use different numerical schemes and time steps with the discharge through the manholes forming the linkage of the models. The source code of SWMM was modified such that the surcharge in the sewer network is represented in terms of hydraulic head rather than overflow volumes. The interacting discharges are determined using weir or orifice equations by taking account of the hydraulic head at manholes and the aboveground water surface for every time step of the sewer network model such that the manhole discharges are treated as point sinks or sources in the 2D model within the same time interval.

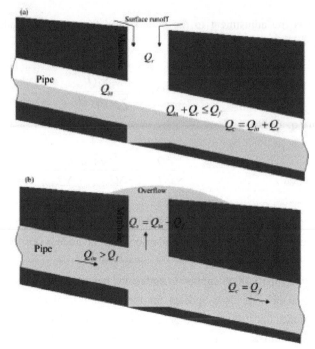

Figure 3-8: Schematic of normal urban drainage and surcharge-induced
inundation, (a) Inflow discharge below the conduit
capacity, (b) Inflow discharge greater than the conduit
capacity

The SWMM model only consists of the sewer network with connecting nodes such as
manholes, basins and outlet. The 2D model consists of the urban surface with a specified
grid for computation, the sources of the flood water and the sinks to the sewer system.
The models are executed separately and linked by exchanging information at the
connecting nodes at each time step of the sewer network model. In this study, the grids
containing the manholes are considered as the locations where the interactions occur. The
time step used in SWMM is also regarded as determining the timing for model linkage.

Interacting discharge
The bidirectional interacting discharge at a manhole is calculated according to the water
level difference between the sewer network node and the aboveground surface. In this
approach, water is assumed to pond in the manhole during flooding, and therefore the
water level in the manhole can be greater than the water elevation in the overland surface.
The upstream and downstream levels for determining the discharge are defined as

$h_U = max\{h_{mh}, h_{2D}\}$ and $h_D = min\{h_{mh}, h_{2D}\}$ respectively, where h_{mh} is the hydraulic head [m] at manhole and h_{2D} is the water surface elevation [m] on the overland grid.

The ground levels of the sewer network nodes reflect the point values of the nodes whereas the grid cell elevations in the 2D model are usually computed as the mean elevation of the topography included within each grid cell. Inconsistency between the ground levels of the sewer network nodes and the 2D grid cell elevations is often present when the connecting nodes are located at local peaks or depressions inside the grids. This inconsistency causes inaccuracies in calculating the interacting discharges and makes the model result sensitive to the DEM. To minimize such inaccuracies the crest elevation Z_{crest} of a manhole is assumed to be equal to the elevation of the grid cell where the manhole is located.

The interacting discharges are calculated using the equations for a free weir, submerged weir and orifice depending on the upstream and downstream water levels (Chen *et al.*, 2007). The free weir equation is adopted when the crest elevation Z_{crest} is between the values of the upstream water level h_U and the downstream water level h_D as shown in Figure 3-9. The discharge is calculated using Eq.(3.29):

$$Q = (h_{2d} - h_{mh}) C_w W \sqrt{2g} (h_{2d} - Z_{crest})^{3/2} \qquad (3.29)$$

If both the upstream and downstream water levels are above the crest elevation of the manhole then either the submerged weir or the orifice equation is used to calculate the interacting discharge. If the upstream water depth above the manhole crest is greater than the area of manhole divided by the weir crest width $(h_U - Z_{crest}) < A_{mh} / W$ then the submerged weir equation (Eq. (3.30)) is used; otherwise the manhole is considered fully submerged and the orifice equation is used (Eq. (3.31)):

$$Q = sign[h_{mh} - h_{2D}] C_w W \sqrt{2g} (h_U - h_{crest})(h_U - h_D)^{1/2} \qquad (3.30)$$

$$Q = sign[h_{mh} - h_{2D}] C_o A_{mh} \sqrt{2g} (h_U - h_D)^{1/2} \qquad (3.31)$$

where C_o is the orifice discharge coefficient, C_w is the weir discharge coefficient and W is the weir crest width. A negative value means flow drains from the surface into the sewer.

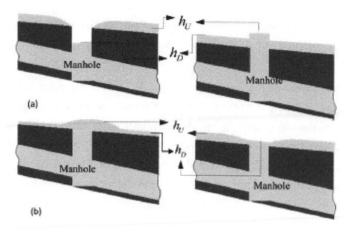

Figure 3-9: (a) Free weir linkage and (b) Submerged weir or orifice linkage

Time synchronisation

The sewer network model (SWMM5) and the overland surface flow routing model use different time step. As Chen et al, 2007 state, the timing synchronization becomes an important issue for connecting both models appropriately, in particular, when variable time steps are used. The models exchange information at the same time. The time step in SWMM5 is made to be fixed while the time step in the 2D model is kept variable. Therefore there is a need to restrict the time step of the 2D model at the time just before the synchronization time to the value given in equation (Eq. (3.32)).

$$\Delta t_{2D_{m+1}} = min\left\{\left(T_{syn} + \Delta t_{1D} - \sum_{i=1}^{m}\Delta t_{2D_i}\right), \Delta t_{2D_{m+1}}^{*}\right\} \qquad (3.32)$$

where $\Delta t_{2D_{m+1}}$ is the time step size [s] used for the $m+1^{th}$ step; T_{syn} is the time of the previous synchronization [s]; $\sum_{i=1}^{m}\Delta t_{2D_i}$ is the total duration of the time step [s] after m step of the 2D model computation following the last synchronization; Δt_{1D} is the time step used in SWMM5 and $\Delta t_{2D_{m+1}}^{*}$ is the time step duration determined by the 2D model for the $m+1^{th}$ step.

Time synchronization between the sewer network model and the overland surface flow model is shown in Figure 3-10.

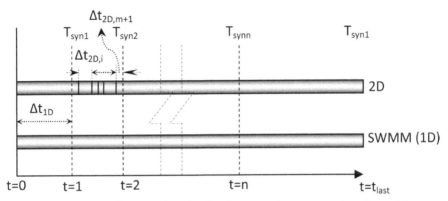

Figure 3-10: Time synchronization between the sewer network model
and the overland surface flow model

3.5 Incorporating rainfall-runoff and the infiltration component in the 2D model

If the required data is available, rainfall-runoff and infiltration processes can be incorporated in the 2D surface runoff model. The type of data required include, the rainfall rate, the initial and equilibrium infiltration capacities (f_0 and f_c in Eq. (2.30)) and the rate of decline of the infiltration capacity of the 2D grid domain.

Initial soil water content appropriate to antecedent conditions should be specified. The time corresponding to any specified water storage, is calculated using Eq. (2.36).

The right hand side of the continuity equation of the 2D model (Eq.(3.1)) would be replaced by q, where q is the vertical flow rate computed from the rainfall rate, the infiltration capacity and the drainage (percolation) rate.

As it is discussed in the literature review, the actual depth of infiltration accumulated during a time step depends on the magnitude of infiltration at the beginning and end of the time step relative to the rainfall rate. The potential infiltration rate f_t at the beginning of a time step can be calculated using Horton's equation (Eq. (2.30)).

There are three possible case in relation to the rate of rainfall P and potential infiltration rate f_t. It should be noted here that the rate of rainfall P represents the rainfall rate and available depth of water for the time step considered.

Case 1: $P \geq f_t$. The depth of infiltration $\overline{f}$ in mm infiltrated between time t and $t+\Delta t$ is given by Eq. (2.37).

Case 2: $f_{t+\Delta t} < P < f_t$. The actual depth of infiltration will be less than the potential infiltration depth. The depth of infiltration during the time interval is computed by Eq. (2.40).

In the above two cases, the vertical flow rate for the same interval is then computed as;

$$q = \left(P - \frac{\overline{f}}{\Delta t} \right) * 1000 * \Delta x \Delta y \qquad (3.33)$$

Case 3: $P \leq f_{t+\Delta t}$. In this case the depth of infiltration is equal to the rainfall rate and therefore q, the vertical flow rate will be zero.

The percolation rate at the end of each time step is calculated using Eq. (2.43) and the change in soil water storage and the soil water storage for the next time step are calculated using Eq. (2.44) and Eq. (2.45) respectively.

Since the 2D employs the ADI procedure to solve the governing equations for flood flows, the q's are evaluated at each half time step.

3.6 Modelling system implementation environment

The modelling systems is developed and implemented in Visual C++ Development Environment of Microsoft visual studio 2008. The development environment provides powerful support for source code editing, source code browsing, and debugging tools. This environment also supports IntelliSense, which makes informed, context-sensitive suggestions as code is being authored (Microsoft). C++ is a general purpose language in computer programming. It is a middle level language that can be used for several purposes in the computer industry.

The current version of the modelling system is a Win32console application with inputs in text files and output in NetCDF file format (Unidata Program Center) that support the creation, access, and sharing of array-oriented scientific data.

3.7 Conclusion

This chapter presents detailed description of models and approaches developed in this research for simulation of urban flood. Details of a newly developed 2D surface flow model are presented. These include; the assumption and simplifications used to arrive at the equations employed in the 2D surface flow model, the numerical methods implemented, the discretization of the governing equations and the procedure used to solve the equations. A methodology to improve flood forecasts in geometrically complex urban environment using coarse grid models is also developed. The methodology involves reinterpretation of the original equations used to develop the 2D surface flow model. The reinterpretation of the equations and use of the volume-depth and flow-are-depth relationships that can be extracted from available fine grid DTM are discussed. Theoretical background and the method used to develop a coupled model which implements a bi-directional linking between the 2D surface flow model and a 1D sewer network model is presented. Finally an algorithm to incorporate infiltration process in the 2D model is discussed. The applicability and success of the models and the methods developed and presented in this chapter will be investigated and measured by the results of the case studies which are presented in the subsequent chapters.

4   2D Model Application to Simulate Surface Flow

4.1 Introduction

This chapter presents case studies to test the application of the 2D urban flood modelling system to urban drainage problems. The 2D modelling system is first tested on hypothetical cases to validate its robustness, and comparisons are made with commercial software which solves the full shallow water equations.

4.2 Application on Flat and sloping plane - Hypothetical Test

A real test of the performance of a 2D surface flow model is to check its performance against flow data collected from various laboratory scale models. However, in the absence of such laboratory test results, the robustness of the 2D model can be tested against established empirical or theoretical models. One such test is to see if the model can produce the normal depth flow speed for a wave front propagating at near steady flow down a sloping plane. The first such idealized case study was performed for a 100m by 1600m sloping plane with a slope of 0.001, a cell size of 2m*2m and a uniform water source across the upper width of the plane. The average wave velocity was calculated after the flow over the plane had reached a steady state condition. The average front wave velocity is simply calculated from the sloping plane test result by dividing the length the front wave travelled by the time taken for the front wave to travel the same distance. This velocity was then compared with the wave velocity which is calculated by Chézy formula as $V = C\sqrt{RS}$, where C is the Chézy roughness coefficient, R is the hydraulic depth, and S is the slope of the plane. The second case study was performed on a flat surface with a single point source having a uniform discharge and observing how the flow propagates on the flat surface.

In the first case, in order to simulate a uniform inflow across the plane water is supplied from manholes located at the center of each cell in the 20th column down the plane. The initial condition is a dry bed and the downstream boundary condition is a rating curve based on uniform flow conditions as defined by the bottom slope and the Chézy coefficient. Test runs were performed for four discharge values and three Chézy roughness coefficients. The time taken for the front wave to travel a distance of 760m from the source was observed from the model results. The water depths from the model results are used to calculate the theoretical velocity according to the Chézy equation and hence the time that it would take for the wave front to travel the same distance. The data collected in this exercise is presented in Table 4-1.

In the flat area case, the source is a single manhole at the center of the area. Figure 4-1 shows the snap shots of the flood propagation for the sloping plane case (a) and the flat area case (b).

4.3   Results of the hypothetical flat and sloping plane case study

The flow propagation along the sloping plane demonstrates that the 2D model produces the same results as the theoretical model. As shown in Figure 4-2 the points are very close to the line of perfect agreement with a coefficient of determination (defined as 1 - (residue variance/corrected total variance)) of 0.99.

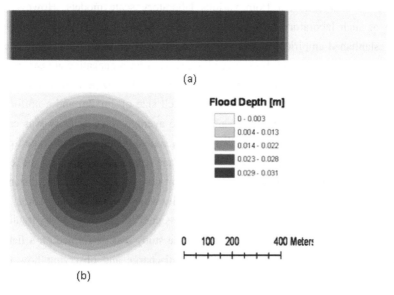

Figure 4-1: Flood propagation on (a) a slopping plane and (b) a flat area
from a manhole at the center

The relationship between travel time and discharge for different roughness values for the sloping plane is shown in Figure 4-3 and for the flat plane in Figure 4-4. Unfortunately there is no readily available empirical or theoretical model for the flat plane case study to validate the result of the 2D model. However, from the figures it can be seen that the relationships both for the sloping plane and the flat plane between travel time and discharge for different roughness values exhibit the same trend.

Table 4-1: Travel time in minutes from the Chézy equation versus travel time calculated from the model results for the wave front to travel a distance of 760m along a sloping plane for different discharges and Chézy roughness values

Discharge [m³/s]	Chézy = 15		Chézy = 30		Chézy = 45	
	Chézy Eq.	Model	Chézy Eq.	Model	Chézy Eq.	Model
0.01	121	123	76	80	57.5	61.7
0.03	84	85	52.8	53.8	40.2	40.7
0.05	71	70	44.5	45	33.8	35
0.1	56	55	35.5	35	27.1	29.6

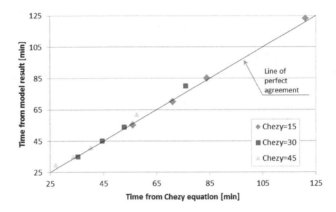

Figure 4-2: A Plot of travel time form Chézy equation versus travel time calculated from model result for the wave front to travel 760m downstream of the source

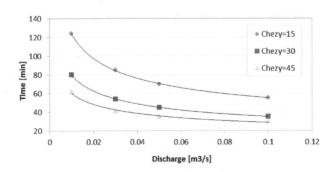

Figure 4-3: Travel time for different discharge values to travel a distance of 760m down the sloping plane with gradient of 0.001

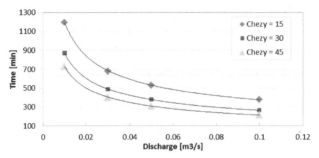

Figure 4-4: Travel time for different discharge values to travel a distance
of 302m on the plane

4.4   Application on a River Valley

The non-convective wave 2D model was used to simulate flood wave propagation down a
river valley following the failure of a dam. This case study was performed using one of
the data sets provided by the UK Environmental Agency for the evaluation of 2D
hydraulic modelling packages (Néelz and Pender, 2010). A test case involving valley
flooding was selected to investigate the performance of the developed 2D model in this
thesis. As a reference the same case study was simulated using MIKE 21 HD Flow
Model, which solves the full shallow water equations (DHI Group, 2009b).

MIKE 21 HD solves the full 2D shallow water equation. In-depth description of the
equations and numerical formulation of model are given in (DHI Group, 2009b). The
model makes use of the Alternative Direction Implicit technique to integrate the equations
for mass and momentum conservation in space-time domain. MIKE 21 HD simulates the
water level variation and flow in response to a variety of forcing functions in lakes,
estuaries, bays and coastal areas. The water levels and flows are resolved on either a
rectilinear grid, a curvilinear grid, a triangular element mesh or any combination hereof
covering the area of interest. It is a general hydraulic model that can be set up to describe
specific hydraulic phenomena. Examples of such applications are tidal exchange and
currents, storm surge, dam-break and urban flooding.

The case study was proposed to simulate flood wave propagation down a river valley
following the failure of a dam, represented by a skewed trapezoidal inflow hydrograph
with a short early peak at 3000m3/s. The valley digital elevation model (DEM) is
approximately 0.8km by 17km and the valley slopes downstream with a slope of

approximately 0.01 in its upper region, easing to approximately 0.001 in its lower region. The inflow hydrograph is applied as a boundary condition along a 260m long line at the upstream boundary, and is designed to account for the typical failure of a small embankment dam and to ensure that both super-critical and sub-critical flows will occur in different parts of the flow field (Néelz and Pender, 2010). A digital elevation model is used, with a prescribed cross-section along the center line. The location of the output points are shown in Figure 4-5 and the inflow hydrograph is shown in Figure 4-6.

All boundaries except the inflow boundary (on the left side of the modelled area) are closed. The model grid resolution used in both models is 50m. A uniform Manning roughness coefficient of n=0.04 is applied everywhere. The models are run until the time T = 30 hours in order to allow the flood to settle in the lower parts of the valley. The water levels and velocity series for seven output points (see Figure 4-5) are shown in Figure 4-7 to Figure 4-10, cross-sections of the peak water levels and the peak velocities along the valley center line are shown in Figure 4-11 and Figure 4-12.

4.5   Results of the hypothetical river valley case study

The two models predict very similar results in terms of peak water levels and peak velocities as can be seen in Figure 4-7 to Figure 4-10. The maximum discrepancies between the two models were almost 0.22 cm which is about 7% of the predicted peak depth at most of the locations which are in excess of 3m for water levels and 0.33 m/s for velocities at point 5.

Prediction of flood arrival time is consistent for all points except point 5 and 6. At point 5, which is located in an approximately 2.5km^2 large pond at the downstream end of the valley where the water finally settles after filling any depressions located further upstream, the arrival time is delayed by approximately 7.5 minutes and the final water level difference is in the range of 0.1m. The delay in the arrival time at point 5 is actually not significant compared to a travel time of almost 3 hours since the start of the flood. Also at point 6, located close to point 1 but at higher elevation, the discrepancy in the peak water level is relatively higher than point 1 with peak level difference of close to 0.1m.

The discrepancies in velocity, calculated as discharge divided by flow area, are relatively higher than discrepancies in peak water levels. The peak velocities are over estimated by the non-convective acceleration wave model at all the points except point 6, and smaller oscillations in the peak flows are observed at points 4, 6 and 7. The peak water levels discrepancies for example vary from 0.2 m/s at appoint 2 to 0.33 m/s at point 5.

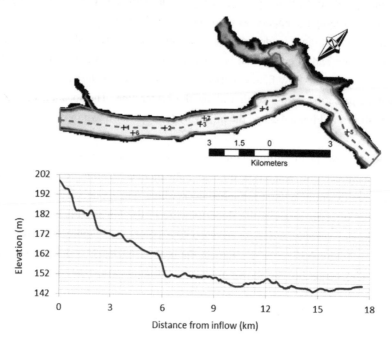

Figure 4-5: DEM of the hypothetical river valley used, with cross-section along the centre line, and location of the output points

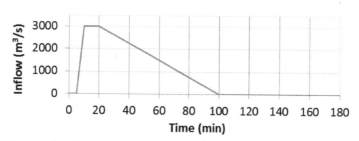

Figure 4-6: Inflow hydrograph used in the test

In terms of peak levels the non-convective acceleration wave model predictions are consistent with the reference model (Figure 4-11 and Figure 4-12). The difference in magnitude is very small. The peak velocity predictions however show marked differences, with more oscillations in the lower part of the valley and slightly higher peak velocities.

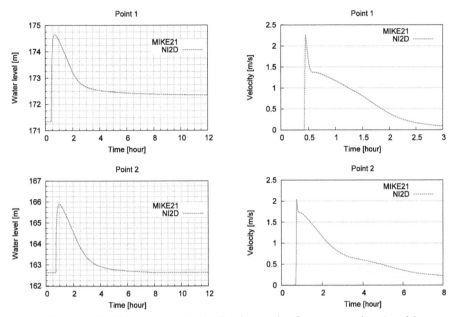

Figure 4-7: Water level and velocity time series for output points 1 and 2

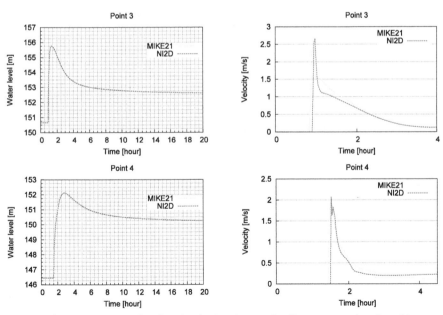

Figure 4-8: Water level and velocity time series for output points 3 and4

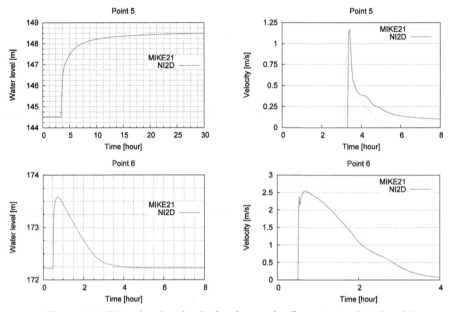

Figure 4-9: Water level and velocity time series for output points 5 and 6

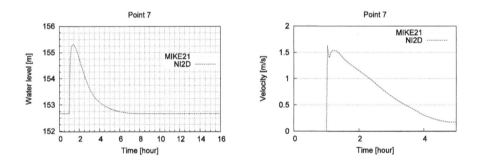

Figure 4-10: Water level and velocity time series for output points 7

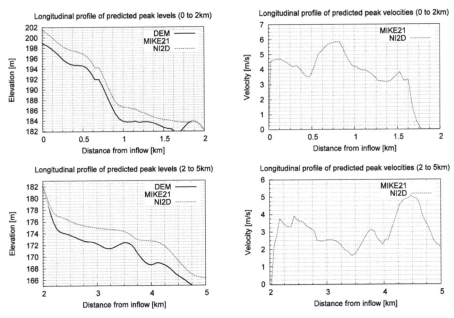

Figure 4-11: Longitudinal profile of peak water levels and peak velocities
along the valley center from 0 to 5km

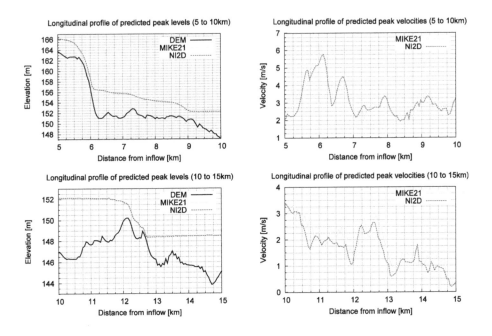

Figure 4-12: Longitudinal profile of peak water levels and peak
velocities along the valley center from 5 to 15 km

4.6   Conclusions on 2D Model Application to Simulate Surface Flow

In this chapter application of the developed non-convective wave 2D model in two case studies is presented. The first test demonstrates that the results of the model for the average front wave velocity are in good agreement with the results obtained from the empirical Chézy equation. The non-convective acceleration 2D model is believed to be inferior to models which employ the full shallow water equations because the convective momentum term is removed. Despite this limitation, the main advantage of this type of model is that it provides relatively good results in cases where models which include the convective momentum term become excessively computationally intensive. In addition, the non-convective acceleration model is proven to be stable and flexible.

The second set of tests has demonstrated that despite the removal of the convective acceleration terms the results of the developed model are very close to the results of the reference model which implements the full shallow water equations. This shows that the model developed in this thesis, whilst neglecting convective acceleration terms which leads to local inaccuracies mainly for velocity predictions, are capable of predicting flood inundation depths successfully.

5   Network and Surface Flow Interaction: 1D–2D Model Application

5.1   Introduction

This chapter is devoted to case studies carried out to simulate the complex interaction between a one dimensional sewer network and the two dimensional above ground surface flow in urban areas. Two case studies are carried out, namely in Dhaka and the Bangkok. The descriptions of these case studies, the inputs used and the simulation results are presented and discussed. Finally, conclusions are drawn from the results.

5.2   Dhaka case study

5.2.1   Dhaka City

The following description of Dhaka city and Segunbagicha catchment is mainly taken from (Ahmed, 2008). Dhaka city, the capital of Bangladesh, is located at the central region surrounded by Buriganga River at the south, Turag River at the west and Balu River at the east. The elevation of the Greater Dhaka area varies between 2 to 13 meter and the urban centre elevation varies from 6 to 8 meter above mean sea level.

High population growth rate along with increased rate of urbanization has made Dhaka the 11th largest mega city in the world. The present population of Dhaka has already exceeded 10 million.

Dhaka has a hot and humid tropical climate. The temperature varies from 18 °C in January and 29 °C in August with an annual average temperature of around 25 °C. The climate of Dhaka can be classified broadly into three seasons: pre-monsoon, monsoon and post-monsoon or dry season. There are thunderstorms accompanied by some rainfall in the pre monsoon period. The months between June to September is monsoon season with 80 per cent of the rainfall occurring this time while it remains cool and dry for the rest of the months. Annually average rainfall of Dhaka is 2000 mm and the average evaporation ranges from 80-130 mm.

Drainage of Dhaka city is managed through two separate sewer systems: one for drainage of domestic wastewater and the other for drainage of storm water. The storm water of

Dhaka City is discharged to the surrounding rivers. Dhaka Water Supply and Sewerage Authority (DWASA) is responsible for the water supply and drainage of Dhaka city.

5.2.2   Segunbagicha catchment

Segunbagicha catchment with a catchment area of 8.3 square kilometer includes the most important business and government office areas of Dhaka City. Floods caused by intense local rainfall occur in the built-up areas of the city several times a year.

Segunbagicha catchment includes areas like Shantinagar, Purana Paltan, Rajarbag comprising some of the most important land uses of Dhaka city such as the commercial hub and centre of business activities of the city and a mixture of residential, commercial uses, government and non-government offices, educational institutes and some small scale industries. The catchment is one of the areas which are frequently faced with flooding problems and some studies were carried out in this catchment to deal with the problem.

5.2.3   Coupled model for Segunbagicha catchment

A coupled model including the drainage system for the Segunbagicha catchment and the associated floodplain has been built by connecting the 2D model developed in this thesis and SWMM5. The drainage network consists of 88 sewer links with a total length of 13,635 meters. Out of this 75 links are circular pipes with a total length of 11,308 meters and 13 links are box culverts with total length of 2,327 meters. The circular sewer pipe diameters range from 450 to 5,500mm and the box culverts sizes are between 2.5 by 2 meters to 5.5 by 4.3 meters. The slopes of the pipes range from 0 to 10%. The underground drainage system consists of circular pipes, box culverts, basins and pumps. Stormwater from sub-catchments is drained by sewer pipes to two basins from which sewage is pumped to Tongi Khal river system (Ahmed, 2008). The layout of the stormwater drainage system in Segunbagicha area is shown as a background in **Figure 5-2**. A MOUSE model of the 1D sewer network was obtained from a previous study (Ahmed, 2008) and the input data was prepared for a SWMM5 model. A 10m DTM including the road network was used to set up the 2D non-convective wave model. In the coupled model the crest elevations of the manholes are assumed to be the same as the grid cell elevations where the manholes are located. Elevations of the manholes are extracted from the DTM and the ground levels of the manholes are then defined. The time series measurements used for the calibration of MOUSE model in the original study were not available for the present study. Hence, the sub-catchment parameters of SWMM5 model were 'calibrated' against the sub-catchment runoff results of the MOUSE model. In other words, the sub-catchment parameters (width, slope and percentage of imperviousness) were adjusted so that the differences between the peak discharges and total runoff volumes from the sub-catchments of the two models are minimised.

Once the SWMM5 model was 'calibrated' against the MOUSE model, the coupled 1D-2D model was used to simulate flooding for a rainfall event as shown in Figure 5-1, and the results were then compared against these measurements as well as against the results of another coupled model built by linking 1D MOUSE and 2D MIKE 21 models. The study of Ahmed (2006) indicates that this rainfall event is equivalent to a 50year average recurrence interval (ARI) event for Dhaka city.

5.3   Results of Dhaka case study

The lack of field data causes difficulties for the calibration and validation of coupled models. The coupled model developed here also suffers from a lack of field data to calibrate the model and validate its results. If field data records were available, the results of the coupled models could be assessed through parameters such as flood area extent, location and flood depth and velocity time series comparisons with field observations. In the absence of field records, the coupled model results were compared to the results of the previous study using a coupled MOUSE and 2D MIKE21 model.

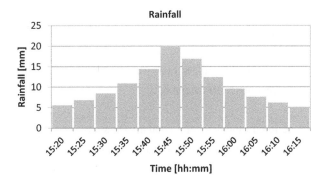

Figure 5-1: A 50 year return period 1 hour rainfall event used in the simulation

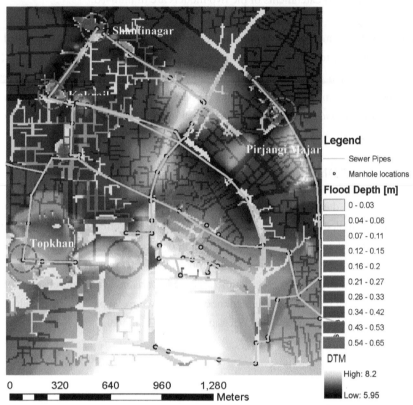

Figure 5-2: Topography and sewer network and maximum flood depth resulting from 5o years return period design rainfall of Segunbagicha catchment

The results of the coupled model for the Dhaka case study were found to be similar to those of the previous study in terms of flooded area extent and flood depths at different locations. The maximum flood depth result of the coupled model for the rainfall event used is shown in Figure 5-2. Flood depth time series comparisons at four locations are shown in Figure 5-3. The results of the SWMM5-non-inertai 2D (SWMM5-NI2D) coupled model are found to be in a good agreement with MOUSE-MIKE21 model results for Shanitnagar and Kakrail location even though the maximum flood depths are slightly underestimated. At Pirjangi Majar the maximum flood depth was reached earlier and was slightly overestimated. Topkhana is located in a relatively higher elevation than the three other locations. The short term surcharge at this location and ponding at the other three locations may be due to that fact that flood flows down to the lower location through road surfaces which act as open channel. The difference in the flood depth time series between the two models may have been due to one or more of the following reasons. For example, although the SWMM5 model was calibrated against the results of MOUSE model, the two models differ the way they handle surcharge flows. MOUSE models surcharge flow using Preissmann slot (DHI Group, 2009a). SWMM5 model uses a different approach as described in section 3.4.1. This may affect the amount of flood overflowing from the sewer network. The different hydrological methods for runoff generation may also contribute in the different result. The purpose of this comparison is to see if the couple model produces comparable results. It is not to compare the two different sewer network modelling systems. However, since the modelled area is relatively flat, non-convective acceleration 2D model can be expected to perform as good as the models which use the full Saint Venant Equations such as MIKE 21 HD.

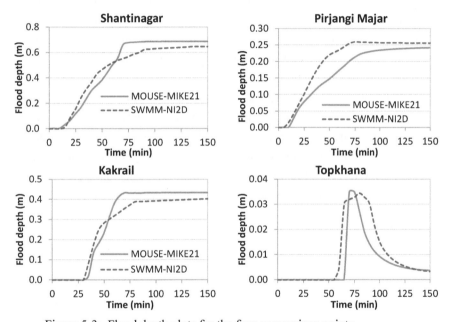

Figure 5-3: Flood depth plots for the four comparison points

5.4 Bangkok Case Study

The coupled 1D-2D model was also used to model the interaction of the sewer network and overland flow for Bangkok. The Bangkok case study concerns a catchment along Sukhumvit Road in the inner part of Bangkok, Thailand, with an area of about 24 square kilometers. The area is characterized by a high density population and heavy urbanisation. Terrain elevations vary from +0.40 meters up to +1.00 meters amsl (above mean sea level). The location of the study area in the larger Bangkok City area is shown in Figure 5-4. The drainage system in the study area consists of closed rectangular channels, circular pipes and pumping stations. This area is well known for its frequent flooding problems, which are mainly attributed to the inadequate sewer system capacity.

The data used in this study was obtained from the Department of Drainage and Sewerage (DDS) of Bangkok Metropolitan Administration (BMA) through the Asian Institute of Technology (AIT). The data includes the following details: geometry of the drainage network, a Digital Terrain Model (DTM) with 5m resolution, time series records of four rainfall events and corresponding time series of water level measurements taken at two locations. The simplified sewer network model contains 207 nodes, 223 links and 193 subcatchments. Critical selection parameters for model simplification were larger conduits and locations of water level stations. The simplification of the model was carried out in such a way that the head losses and storages from manholes and pipes which were found to be outside of the simplified model network were incorporated into manholes of the final model network. The final 1D-2D model layout, the location of the two measurement stations and the DTM of the study area are shown in Figure 5-5.

The rainfall data was collected at four stations and includes two events from 1998 (one from August 16th and one from October 1st) and two events from 2002 (one from October 5th and one from October 7th). For the 1998 events the time series water level measurements were not taken but a reasonable amount of eyewitness data concerning maximum street water levels has been assembled. In 2002, in addition to the collection of rainfall data, time series water level measurements in manholes and streets were taken at two locations. This data was used for calibration of the 1D-2D model. The calibration results are given in Figure 5-6.

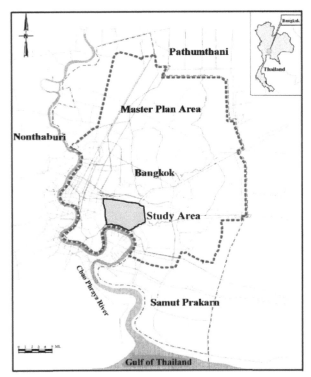

Figure 5-4: Map of Bangkok City and the study area.

Parameters used in the calibration of 1D-2D model include Manning's roughness coefficient of the pipes in the 1D model, the runoff coefficient and the discharge coefficient of the manholes used for calculating the interacting discharges between the 1D and the 2D models. Assuming that the drainage network of the area remained unchanged between 1998 and the 2002, which was confirmed by the authorities, the 1998 rainfall events were used to validate the coupled model. As mentioned above, in 1998 time series measurements were not carried out but there is a considerable amount of eye witness records collected along the streets and compiled by the authorities. Therefore, these records were used in the validation of the 1D-2D model.

5.5   Results of Bangkok case study

Illustrations of calibration results concerning two measurement locations in the Sukhumvit area in Bangkok are given in Figure 5-6. The figure gives a comparison of simulated and measured water levels. The statistical measure used to assess the model calibration is the Coefficient of Determination, and the values obtained were between 0.90 and 0.99; see Table 5-1.

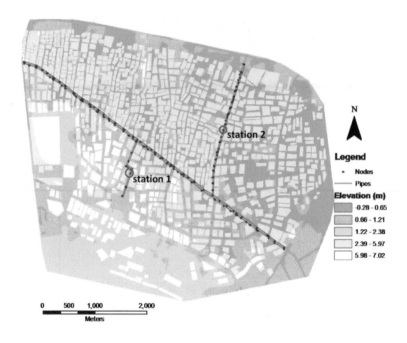

Figure 5-5: Drainage network, location of the measurement stations and topographic map of the study area, Bangkok case study

Table 5-1 : Root Mean Square Error (RMSE) and Coefficient of Determination (R2) of the calibration results at the two measurement locations

| Location | Time of record available | | | |
| | October 5, 2002 from 15:30 to 22:30 | | October 7, 2002 from 17:30 to 18:45 | |
	RMSE (m)	R^2	RMSE (m)	R^2
Station 1	0.073	0.969	0.094	0.996
Station 2	0.028	0.993	0.125	0.923

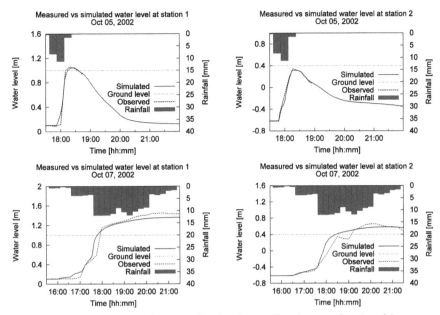

Figure 5-6: Measured versus simulated water levels at station 1 and 2.

The simulated peak food depths for the validation events (i.e., August 18th and October 1st, 1998 rainfall events), and the ranges of indicative flood depths obtained from eye witness records, are given in Table 5-2 and Table 5-3. The maximum flood depth map for the October 1st, 1998 rainfall event is shown in Figure 5-7 for illustrative purpose. The simulated flood depths are found to be close to or within the range of the eye witness records. The results show the capability of the coupled model to simulate the complex interaction between the sewer flow and the surcharge-induced inundation.

Table 5-2: Flood depth comparison for the August 16, 1998 rainfall event

Location	Maximum simulated flood depth (m)	Maximum observed (from eye witness) flood depth (m)
1	0.63	0.50 – 0.70
2	0.24	0.10 – 0.30
3	0.46	0.40 – 0.60
4	0.30	0.30 – 0.50
5	0.55	0.40 – 0.60

Table 5-3: Flood depth comparison for the October 1, 1998 rainfall event

Location	Maximum simulated flood depth (m)	Maximum observed (from eye witness) flood depth (m)
1	0.80	0.70 − 0.90
2	0.38	0.30 − 0.50
3	0.53	0.40 − 0.60
4	-	-
5	0.58	0.60 − 0.80

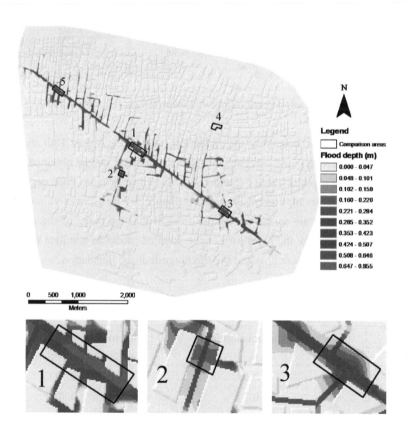

Figure 5-7: Maximum flood depth map for October 1, 1998 rainfall event

5.6   Conclusion

In this chapter the ability of the coupled model to simulate the interaction of below ground sewer network flow and above ground surface flow has been investigated. The objective of developing the coupled model is to assess more adequately the performance of storm or combined sewer systems and their associated surface flooding in order to predict the value of potential flood damages and to provide decision makers with information to design appropriate measures. The couple model was built by coupling SWMM5 model to simulate the flow in sewer network and the non-inertia 2D model to simulate the surface flow. This enables elements which are essentially one-dimensional (e.g., drains, culverts, channels) to be modelled explicitly using a sewer network model (SWMM5 in this case), while overland processes are modelled with a two-dimensional schematization. The results of the two cases studies demonstrate that the coupled model is capable of simulating the interaction between sewer networks and surface flood flow with good efficiency.

5.6. Conclusion

In this chapter, the ability of the coupled model to simulate the interaction of belowground and aboveground flow and aboveground storm flow has been evaluated. The relevant descriptive...

6   Topographic Grid Generalization for Urban flood Modelling

6.1 Introduction

The modified non-convective wave 2D model, which is reformulated to make use of information that can be extracted from the fine resolution grid for coarse grid models (for convenience hereafter called the 'modified model') and the original model, which does not take this kind information into account (hereafter called the 'standard model'), are applied to build coarse grid models for the simulation of urban flooding. The case study and results of the coarse grid models are described and discussed in this chapter.

6.2 The Case Study

Coarse grid models are built and applied to simulate shallow inundation originating from an inflow to an area. The test domain to be modelled consists of a rectangle 0.4 km by 0.96 km, with dense urban development either side of two main streets and a topologically complex network of minor roads. The area is a mix of some steep sections of road and local depressions where water may pond. To characterise the topography and topology for this study site, a 0.5m bare earth DEM obtained from UK Environmental Agency was aggregated to a 2m resolution bare earth DEM and buildings, kerbs and roads were all reinserted, based on their locations in the digital map layer. The grid cells covered by the buildings were raised in elevation by 6m to represent building height. The precise height value is unimportant here as flow depths during all simulations were always less than two meters and the purpose of this processing step is to allow buildings to be represented as 'islands' that water must flow around. The result is a high-resolution DEM (hereafter termed the benchmark DEM) shown in Figure 6-1 with a realistic representation of urban morphologic features. A land-cover dependent roughness value is applied, with two categories: roads and pavements with Chézy roughness value of 45 and any other land cover type with Chézy roughness value of 20.

Flooding at the site is caused by an inflow that enters near the north-east corner of the domain. The inflow boundary condition used in this study consists of the inflow hydrograph shown in Figure 6-2, which is imposed as an 8m long inflow boundary shown by the red line in Figure 6-1.

101

Legend

High : 42.6

Low : 21.07

Figure 6-1: 2m by 2m resolution Benchmark DEM for the hypothetical case study

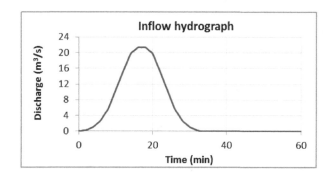

Figure 6-2: Inflow hydrograph used in the hypothetical case study

The inflow lasts less than 60 minutes, but each simulation was continued for 60 minutes to allow water to come to rest and pond in depressions. After 60 min the hydraulic part of the event is effectively over and water levels have ceased to change significantly, although a considerable volume of water remains in the model domain. All external boundaries for each model were closed with zero mass flux.

6.3   Generating coarse grid data set

The topographical dataset for building the coarse models is generated in two ways. Due to the large and often complex terrain surface areas associated with urban flood models, the generalising method that is usually chosen is one which requires minimal processing power, is not computationally demanding and which can be applied to larger areas. The common practice of generalizing a regular topographical grid which satisfies the above requirements is to group neighbouring cells together and taking average values for the coarse cells. For the standard model, a 4m by 4m resolution coarse grid is prepared from the 2m by 2m grid by grouping 2 cells by 2 cells into one cell as shown in Figure 6-3 below.

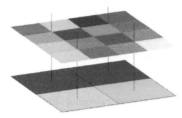

Figure 6-3: Generalized grid using a 2x2 averaging window.

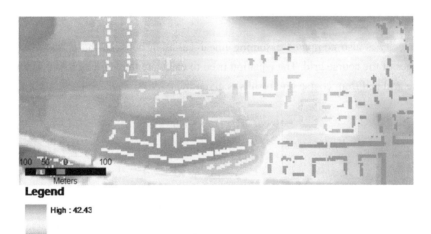

Figure 6-4: 4m by 4m resolution generalized DEM from the 2m by 2m grid DEM

A 4m by 4m resolution coarse grid of the standard model and 4m by 4m, 8m by 8m and 16m by 16m coarse grids of the modified model are used to simulate the flooding condition for the case described above and the results are compared with the fine grid resolution (the benchmark) model results.

The modified model uses the fine grid DEM to produce the volume - depth and flow-area-depth relationships. The model results, including water depths at cell centers and velocities at cell boundaries for the coarse model, are at different resolutions than the benchmark model result. These results can be converted back into the benchmark model resolution using equations (6.1) and (6.2). This is logical in the modified model case because the results are computed taking into account the volume-depth and the flow-area-depth relationships derived from the fine grid resolution.

The fine grid water depth is computed from the coarse grid water depth results as the water level of the course grid minus the surface elevation of the fine grid.

$$Wd_f(n,m) = \max\left[WL_c(i,j) - SE_f(n,m), 0\right] \qquad (6.1)$$

where $WL_c(i,j)$ is water level of coarse grid cell (i,j)

$\quad\quad Wd_f(n,m)$ is water depth of the fine grid cell (n,m)

$\quad\quad SE_f(n,m)$ is surface elevation of the fine grid cell (n,m) with in

the coarse grid cell

The velocity is also computed assuming linear variation between the entrance and exit velocities of the course grid. The equation used to calculate the velocity in the x-direction for fine grid cell from coarse grid results is given by.

$$V_{fx}(n,m) = \begin{cases} 0 & \text{if } Wd_f(n,m) = 0 \\ V_{cx}(i,j) + \left(V_{cx}(i+1,j) - V_{cx}(i,j)\right) * k * \dfrac{DY_f}{DY_c} \end{cases} \qquad (6.2)$$

where $V_{fx}(n,m)$ is the left side velocity for fine grid cell with in the coarse grid

Cell (i,j),

$\quad V_{cx}(i,j)$ and $V_{cx}(i+1,j)$ are the coarse grid cell left and right side velocities respectively for coarse grid (i,j)

$\quad k$ is fine grid cell counter in a single row within the coarse grid (varies from 0 to n where n is the number of fine grid columns contained in the coarse grid,

$\quad DY_f$ and DY_c fine grid and coarse grid cell sizes in the y-direction respectively.

Figure 6-5 depicts how coarse grids model results are converted back to fine grid results.

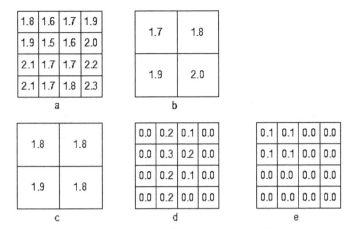

Figure 6-5: (a) Fine grid DEM, (b) coarse grid DEM, (c) water level
from coarse grid, (d) fine grid water depth computed from
the coarse model result for the modified model and (e) for
the standard model

6.4　Model Results

The model results are compared in terms of the maximum flood depth that occurred
during the simulation period, the flood depth and velocity time series at six selected
locations and the computational time required to finish the simulation. The maximum
flood extent and flood depth are shown in Figure 6-6 and the water level and velocity
time series for the six comparison points are shown in Figure 6-7 and Figure 6-8.

The Coefficient of determination and Nash-Sutcliffe efficiency are computed for the
flood depth and velocity series of the results from the two 4m by 4m coarse grid models
and are shown on Table 6-2.

The Coefficient of determination R^2 is defined as the squared value of the coefficient of
correlation. It is calculated as:

$$R^2 = \left(\frac{\sum_{i=1}^{n}(O_i - \bar{O})(P_i - \bar{P})}{\sqrt{\sum_{i=1}^{n}(O_i - \bar{O})^2}\sqrt{\sum_{i=1}^{n}(P_i - \bar{P})^2}} \right)^2 \quad\quad (6.3)$$

with O observed and P predicted values.

In this case, the benchmark model results are considered as observed values and the coarse model results are considered as predicted values. The Coefficient of determination estimates the combined dispersion against the single dispersion of the observed and predicted series. The range of R^2 lies between 0 and 1 which describes how much of the observed dispersion is explained by the prediction. A value of zero means no correlation at all whereas a value of 1 means that the dispersion of the prediction is equal to that of the observation.

The Nash-Sutcliffe efficiency E proposed by Nash and Sutcliffe (1970) is defined as one minus the sum of the absolute squared differences between the predicted and observed values normalized by the variance of the observed values during the period under investigation. It is calculated as:

$$E = 1 - \frac{\sum_{i=1}^{n}(O_i - P_i)^2}{\sum_{i=1}^{n}(O_i - \overline{O})^2} \qquad (6.4)$$

The range of E lies between 1.0 (perfect fit) and $-\infty$. An efficiency of lower than zero indicates that the mean value of the observed time series would have been a better predictor than the model.

The sums of the roots of the squared errors for the maximum flood depths in the coarse grid models are computed to quantify the difference in the maximum flood depths results between the coarse models and the benchmark model. The errors are computed as the cell by cell difference of the maximum flood depth and the sum of the roots of the squared errors are computed using Eq. (6.5).

$$SRSE = \sum_{i=1}^{N_x}\sum_{j=1}^{N_y}\sqrt{(Error(i, j))^2} \qquad (6.5)$$

where Nx and Ny are the number of columns and rows of the grid.

Time of computation
An advantage of the coarse grid models is to reduce the computational time required in 2D modelling. The up-sizing of fine resolution data to coarse resolution data leads to a loss of accuracy in the model results. This leads to a trade-off between model result

accuracy and computational time. The computational times required to finish the simulation of the benchmark model and the coarse models for the case study are shown in Table 6-1. These computational times are recorded for a Windows 7 based PC with Intel Core i5 CPU and 4GB RAM.

Table 6-1: Computational time for the fine resolution and coarse grid models

Grid resolution (m*m)	Model type	Computation time (min)
2*2	Standard	40.5
4*4	Standard	6.95
4*4	Modified	8.9
8*8	Modified	1.62
16*16	Modified	0.75

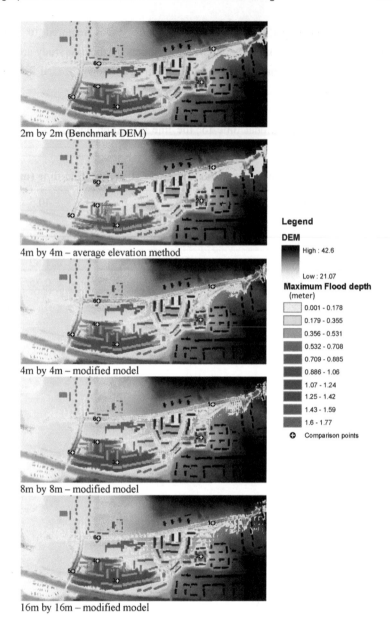

2m by 2m (Benchmark DEM)

4m by 4m – average elevation method

4m by 4m – modified model

8m by 8m – modified model

16m by 16m – modified model

Figure 6-6: Maximum flood depth for the fine and coarse grid models

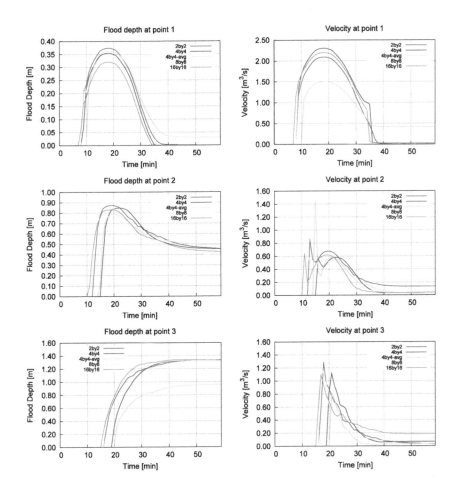

Figure 6-7: Flood depth and velocity time series for point 1, 2 and 3

The performance of the coarse model compared with that of the fine resolution (the benchmark model) indicates that the results of the modified model are significantly better than the results of the coarse model, which does not take into account any of the information that can be extracted from the fine resolution DEM. As it is seen in the maximum flood depth plots in Figure 6-6, the flood depth and velocity time series plots in Figure 6-7 and Figure 6-8 and the flood depth error plots in Figure 6-9 the coarse model results of the modified model which make use of the volume-depth and flow-area-depth relationships derived from the fine resolution DEM are much closer to the benchmark model results in terms of flood depth, inundation extent and time of flooding.

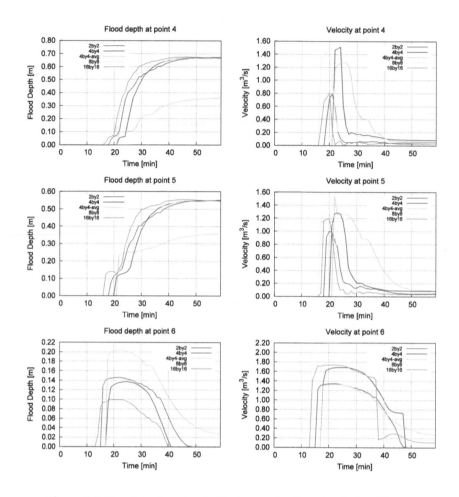

Figure 6-8: Flood depth and velocity time series for point 4, 5 and 6.

The comparison of the 4m by 4m coarse model for the two types models in terms of Coefficient of determination R^2 and Nash-Sutcliffe efficiency criteria E is shown in Table 6-2 for the flood depth and velocity time series at six comparison points. From the table it can be seen clearly that the modified model prediction are in good agreement with the benchmark model with a minimum R^2 of 0.909 and E of 0.871 for the flood depth and a minim R^2 of 0.575 and E of -2.685 for the velocity, whereas the standard model has a minimum R^2 of 0.724 and E of 0.126 for the flood depth and a minimum R^2 of 0.030 and E of -6.197 for the velocity. The velocity time series prediction by the coarse models of the modified model, though better than that of the standard model, are not in good agreement with the fine resolution model.

Table 6-2: Comparison between the two methods for the 4m by 4m coarse grid models

	Flood depth				Velocity			
	Modified method		Standard method		Modified method		Standard method	
	R^2	E	R^2	E	R^2	E	R^2	E
Point 1	0.985	0.968	0.908	0.906	0.989	0.984	0.791	0.692
Point 2	0.999	0.996	0.724	0.678	0.972	0.914	0.495	0.204
Point 3	0.999	0.999	0.803	0.449	0.940	0.889	0.264	0.146
Point 4	0.996	0.996	0.812	0.126	0.572	-2.685	0.030	-6.197
Point 5	0.995	0.994	0.790	0.475	0.805	0.159	0.116	-3.104
Point 6	0.909	0.871	0.916	0.768	0.792	0.456	0.769	0.766

The 8m by 8m and the 16m by 16m coarse models of the modified model also perform better than the 4m by 4m coarse model of the standard model in all measures including error in maximum flood depth (see Figure 6-9) and time series of flood depth and velocity (see Table 6-3).

The significant advantage of the coarse models is evident in terms of computational time. The computational time was reduced from almost 41 minutes for the benchmark model (2m by 2m resolution) to less than 1 minute for 16m by 16m resolution model as shown in Table 6-1. This reduction of computational time is significant particularly for the modified model as the loss of accuracy is a minimum compared to the standard coarse model.

Table 6-3: R^2 and E for the 8m by 8m and 16m by 16m coarse grid models of the modified model

	8by8m				16by16m			
	Flood depth		Velocity		Flood depth		Velocity	
	R^2	E	R^2	E	R^2	E	R^2	E
Point 1	0.968	0.938	0.974	0.966	0.956	0.886	0.959	0.951
Point 2	0.940	0.919	0.719	0.561	0.943	0.938	0.682	0.540
Point 3	0.911	0.910	0.845	0.844	0.982	0.972	0.733	0.660
Point 4	0.909	0.910	0.418	-2.721	0.982	0.972	0.306	0.086
Point 5	0.902	0.902	0.572	0.069	0.978	0.967	0.584	0.433
Point 6	0.730	0.649	0.793	0.334	0.774	0.649	0.774	0.544

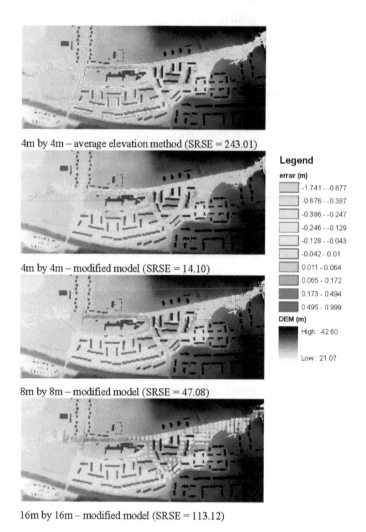

4m by 4m – average elevation method (SRSE = 243.01)

4m by 4m – modified model (SRSE = 14.10)

8m by 8m – modified model (SRSE = 47.08)

16m by 16m – modified model (SRSE = 113.12)

Figure 6-9: Error in maximum flood depth occurred throughout the simulation for the course grid models

As it is evident in the result, grid coarsening reduces the computational time, however the reduction in computational time is not a factor of the grid size change. This would have been the case if the model used a content time step. But because of the fact that the model uses adaptable time steps depending on the convergence tolerance of the solution, the computational time reduction may not be a factor of the grid size change during coarsening.

6.5 Conclusion

Urban environments consist of fine scale features which affect the flow path of surface flow. These features are very important in representing the characteristics of urban flood modelling. A fine resolution DEM represents these complex features of urban geometry well. However, the use of a fine resolution DEM in urban flood modelling is computationally intensive and therefore time consuming for large scale application. This has created the need to find methodologies which will help to reduce computational time required for more efficient application of two dimensional models for urban flood modelling. One viable solution is the use of a coarse spatial resolution DEM. However, the coarsening or generalisation of DEM within urbanised areas can lead to significant changes to the topology and resulting 2D surface flow. Even relatively small changes in model resolution have considerable effects on the predicted inundation extent and the timing of flooding. A reduction in the resolution of the geometry leading to the construction of coarse models for urban flood modelling has a significant effect on flood propagation in urbanised areas because the geometry of urban areas tends to be highly detailed and generalization can lead to the diffusion and subsequent loss of urban features such as curbs, channels and buildings which affect the flow paths of any surface flow.

In this chapter an application of a methodology developed to enable efficient model application at coarse spatial resolution, while retaining important information that can be extracted from detailed (fine) spatial resolution available, is carried out. To retain and use important information of complex urban geometry in coarse resolution models or, in other words, to capture small scale urban features of the fine resolution DEM in the coarse resolution DEMs, volume-depth and flow-area-depth relationships were extracted from fine resolution DEM and used in the modified 2D urban flood model. As it is evident in the case study described and discussed in this chapter, such an approach enables the use of coarse resolution models for urban flood modelling without a significant loss of accuracy and with the advantage of reducing the computational time. From the case study, it can be concluded that;

- The prediction of flood extents by both coarse models conforms with the fine resolution model,
- Flood depth and velocity time series predictions by the standard coarse model are not consistent with the fine grid model (the predictions show time delays and relatively poor results for water depths, specially for output points 3, 4 and 5),
- The coarse models of the modified model predict the flood depths more accurately,

- The modified model shows good flood depth predictions for 4by4m, 8by8m and 16 by16m grid resolutions with a considerable reduction in computation time and without significant loss of accuracy.
- The results show that flood velocities are more sensitive to model grid resolution than the flood depths.

The modified model approach for coarse grid modelling, using a coarse resolution grid with volume-depth and flow-area-depth relationship extracted from fine resolution grid, seems a most promising method as it enables the use of coarse grid models which help to reduce computational time considerably and yet the influences of small scale urban features on the flow are taken in to account in terms of storage-depth and flow-area-depth relationships.

7   Effect of Infiltration in Urban Flood Modelling

7.1   Introduction

The rainfall-runoff transformation on urban catchments involves various physical processes. These processes include interception and evapotranspiration by the vegetation, infiltration of water in the soil, surface runoff and evaporation of surface and soil water. Rainfall-runoff models for urban catchments usually do not take into account all of these processes.

Modelling of the process of infiltration of surface water into the soil is essential for the proper planning and design of urban drainage systems. With the increased emphasis placed on the use of spatial measures of flood peak control like permeable pavements, detention areas, vegetated resistance, etc., such measures have collectively become known as best management practices (BMPs) in pursuit of what are called Sustainable Urban Drainage Systems (SuDS). Such systems consider the infiltration process in urban flood modelling which gives essential insight to the use of the flood peak control methods in urban storm water management.

The effect of infiltration in urban flooding is normally considered to be not significant because urban areas are assumed to be mostly paved and the infiltration capacity is considered to be a minimum in relation to the rainfall rate which causes the flooding. However, to evaluate the effect of infiltration on urban flooding, an algorithm to simulate infiltration using a modified Horton method is included in the 2D model as discussed in the previous chapter. Such a facility will help to evaluate and assess the effectiveness of different source control techniques and to make recommendations on the selection of preferred types of source control; in particular, what type of source control technique and for what condition.

7.2   Infiltration test

The 2D model with infiltration process is used to simulate rainfall-runoff processes on a 2D grid of different initial infiltration capacity. The model is tested simply if the infiltration process is modelled as expected and to see the effect of infiltration on the runoff generation process.

The first test is carried on a 2m by 2m flat grid with the following infiltration parameters. 50mm/hr initial infiltration capacity, 5mm/hr final infiltration capacity and 4.14/hr decline rate. This grid is subjected to two uniform rainfall intensities as follows.

$$RF(t) = \begin{cases} 100 \ mm \ / \ hr \ if \ t \in [0,10 \ \text{min}] \\ 0 \ mm \ / \ hr \qquad \text{Otherwise} \end{cases}$$

$$RF(t) = \begin{cases} 100 \ mm \ / \ hr \ if \ t \in [0,10 \ \text{min}] \ \& \ [30,40 \ \text{min}] \\ 0 \ mm \ / \ hr \qquad \text{Otherwise} \end{cases}$$

The model result on a single grid is shown in Figure 7-1. The infiltration rates, storage in the soil and flood depth on the surface are plotted against time.

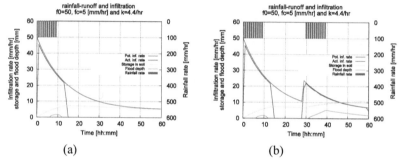

(a)   (b)

Figure 7-1: Infiltration rate, storage and flood depth calculated in a single grid.

After testing the model capability of simulating infiltration process, the same flat grid with maximum and minimum initial infiltration capacity is subjected to the second rainfall intensity as described above. The maximum and minimum initial infiltration capacities are taken from Table 7-1 below which shows the commonly used Horton infiltration parameter values. The results are shown in Figure 7-2 below.

To observe the effect of infiltration capacity on runoff generation, another test was carried out using a hypothetical case of plane with a slope of 2% in one direction of space, as shown in Figure 7-3 (a), with four different infiltration capacities. In the first case, the surface is assumed to be impermeable and hence no infiltration. For the second case, the surface is assumed to have a high infiltration capacity with initial infiltration capacity of 127mm/hr, final infiltration capacity of 5mm/hr and 4.14/hr decline rate and in the third case, the surface is assumed to have minimum infiltration capacity with initial infiltration capacity of 18mm/hr, and the rest is similar to the second case. The fourth case considered a surface with strips of 20m wide maximum and minimum initial infiltration capacities.

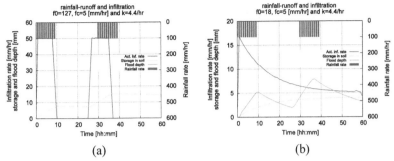

(a)   (b)

Figure 7-2: Infiltration rate, storage and flood depth calculated in a single grid for maximum and minimum infiltration capacity surface

The following lists include commonly used Horton infiltration parameter values(Akan, 1993).

Table 7-1: Commonly used Horton infiltration parameter values (Akan 1993) - Initial infiltration capacity

Soil Type	f_o (mm/hr)
Dry sandy soils with little to no vegetation	127
Dry loam soils with little to no vegetation	76.2
Dry clay soils with little to no vegetation	25.4
Dry sandy soils with dense vegetation	254
Dry loam soils with dense vegetation	152.4
Dry clay soils with dense vegetation	50.8
Moist sandy soils with little to no vegetation	43.18
Moist loam soils with little to no vegetation	25.4
Moist clay soils with little to no vegetation	7.62
Moist sandy soils with dense vegetation	83.82
Moist loam soils with dense vegetation	50.8
Moist clay soils with dense vegetation	17.78

Table 7-2: Commonly used Horton infiltration parameter values (Akan
1993) - final infiltration capacity and decline rate

Soil Type	fc mm/hr	k (1/min) (1/hr)
Clay loam, silty clay loams	0–1.3	0.069 (4.14)
Sandy clay loam	1.3–3.8	0.069 (4.14)
Silt loam, loam	3.8–7.6	0.069 (4.14)
Sand, loamy sand, sandy loams	7.6–11	0.069 (4.14)

$$RF(t) = 100mm / hr\ t \in [0, 60\,min]$$

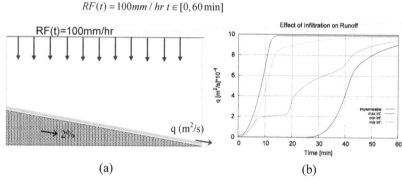

(a)   (b)

Figure 7-3: Effect of infiltration on runoff generation, (a) test
configuration and (b) discharge at downstream for different
infiltration capacity

A uniform rainfall intensity of 100mm/hr is applied for 60minutes and the resulting
discharge downstream is calculated and plotted against time as shown in Figure 7-3(b).

7.3   Results and discussion

The infiltration tests show that the model produced sensible results as one would expect.
In the first test, as the rainfall rate is more than the infiltration capacity, the actual
infiltration rate is almost equal to the potential rate and the storage in the subsurface soil
increased. Flood depth or ponding in the surface is created. Few minutes after the rainfall
stopped, the actual infiltration rate decreased to zero and the storage in the soil gradually
reduced to the final infiltration capacity as some water drain from wet soil to deeper
profile. The actual and potential infiltration rates pick as there is water to infiltrate as
shown in Figure 7-1(b). The infiltration rates increased for the second rainfall event and
start declining as the potential is fulfilled. The storage in the soil increased and water
ponded on the surface. After the rainfall stopped the ponded water depth reduced
gradually.

The second infiltration test was carried out to see the effect of potential infiltration rate on the actual infiltration rate, the storage on the soil and on the flood depth produced on the surface. As shown in Figure 7-2, two rainfall events of 100mm/hr intensity were applied on surfaces of maximum and minimum potential infiltration capacity. For the surface with maximum potential infiltration capacity, the actual infiltration rate increases and decreases with the rainfall. As the potential infiltration rate is more than the rainfall intensity, no surface flood is generated, but the soil water or soil storage increased. For the surface with the minimum initial infiltration rate, the actual infiltration rate decreases gradually and the soil storage increases slowly. Since the rainfall intensity is greater than the infiltration rate, surface flood is generated. The second rainfall event didn't have effect on the actual infiltration rate and the soil water storage as water ponded on the surface from the first rainfall event satisfied the infiltration capacity, however the flood depth increased with the second rainfall event and reduced gradually after the rainfall stopped. These two infiltration test showed that the model can simulate infiltration process pretty reasonably.

The final test shows the effect of infiltration on runoff generated. The result shows surfaces with high initial infiltration capacity reduce and delay the surface runoff generated.

7.4   Conclusion

Generally, for urban flood modelling rainfall-runoff modelling is performed using hydrological models and the results are used as boundary conditions or in the case of coupled 1D-2D models the rainfall-runoff modelling is incorporated in the one dimensional models. The hydrological models in such cases can simulate runoff to a reasonable accuracy on large temporal and special scales however; they are not suitable to simulate flooding for the purpose of computing inundation map, water depth, propagation velocities and arrival time in urban flood plains. Incorporation of rainfall-runoff and infiltration process in the 2D model enables distributed modelling of rainfall-runoff process and simulation of flood events to produce inundation maps, flood arrival time, flood depth and flood hydrographs at any point in time and space in the floodplain.

The above tests show that the procedure incorporated in the 2D model for modelling of infiltration, which is based on a modification of Horton's equation, allows for simulation of the infiltration process within the flood modelling. Infiltration behaviour is expressed as a function of soil water content and the infiltration rate decays with time. This allows for recovery of infiltration capacity during intermittent rainfall. Antecedent conditions

119

can also be taken into account by specifying the soil water content at the start of simulation. Since the model parameters are not calibrated and the results are not validated against measured infiltration data, it is not possible to quantify the performance of the model. However, the performance of the model in simulating infiltration is found to be satisfactory in that it provides sensible results that one would expect.

8   Conclusions and Recommendations

8.1   Conclusion

The impacts of flooding are especially devastating and costly in urban areas as these areas are densely populated and contain vital infrastructures. Urban flood risks and their impacts are expected to increase as urban development in flood prone areas continues while aging drainage infrastructures limit the drainage capacity in existing urban areas. Increasing rain intensity as result of climate change further increases the risk of flooding in various cities around the world. The increased risk and severe consequence of flooding drives the need for the development of cost-effective flood mitigation strategies as part of sound urban flood management plans. This in turn drives the need for efficient prediction of characteristics of flood propagation in urban areas. While flood modelling is a fairly recent practice, an enormous amount of work has been performed on urban flood modelling during the last decade. The advent of powerful computing machines and development of numerous numerical techniques paved the way for use of computer models to pervade in all aspect of water management and more specifically flood propagation. Urban flood modelling attempts to qualitatively describe the characteristics and evolution of flood flows that occurs when a large amount of water moves along drainage systems and urban flood plains. The nature and origin of floods vary ranging from slow reservoir-filling like inundation resulting from long lasting rainfall to flash floods resulting from high intensity rainfall or the failure of a dam or other control structure.

Urban flood models describe the flood characteristics in drainage systems and urban flood plains using mathematical equations numerically embedded into a computer code. One of the most important data items to build inundation models is topographic data which describe elevation and shape of natural and artificial features on the land surface that drive the surface flow. Until relatively recently this important data for building inundation models were generated through a time consuming and expensive process in which contour lines were produced by interpolating spot heights measured manually on site. Because of technological advancement like LiDAR and IFSAR it is now possible to generate detailed topographical data over large areas in a relatively short time frame. The choice of required resolution varies from application to application and it depends on many factors such as the size of the study area, the minimum size of key features required by the model, the complexity of model and the type of model used for simulation. High resolution topographic data which represents individual buildings and other topographic features such as natural pits, ridges, channels, alleys and man-made features, is essential for detailed prediction of flood flows in urban areas. However, hydraulic model simulations on high resolution grids is usually very demanding in terms of computational cost and may make the use of detailed topographic data unfeasible in situation where there is a need to reduce computational time. In such circumstances, topographic data is often generalised to a more manageable resolution and floodplain models are built at much coarser resolutions such that complicated flow patterns due to vegetation, buildings or any other man-made structures become sub-grid scale processes.

However, the generalisation of topographic data within urban environments leads to significant changes in the topography due to the spreading or disregarding of dominant features. As a result, 2D models with a lower resolution more likely produce inaccurate flood simulation results than high resolution 2D models. Several methods are devised in order to keep the information that can be obtained from high resolution topographic data in coarse grid models. One of the novel approaches researched in this thesis involves modifying the equations which describe the flow of water on surface when making the topographic grid coarser. The next paragraphs will summarize the approach taken within this thesis and highlight the main conclusions of the research.

Development of a non-inertia 2D surface flow model

The propagation of a flood over a flood plain is a three dimensional time dependent, incompressible fluid dynamics problem with a free surface. The flow can be considered single phase if erosion and deposition effects are neglected. The Navier–Stokes (NS) equations, which arise from applying Newton's second law to fluid motion, perfectly describe the dynamics of flood wave propagation. However, the solution of the 3D NS equations is difficult because of presence of turbulence and huge difference of length and time scales. Therefore it cannot be applied to simulate practical cases of flooding. Considering a specific problem such as shallow water flows in which the horizontal scale is much larger than the vertical one, the Shallow Water Equations (SWE), which can be derived from the Navier-Stokes equations by integrating them over the depth of flow, will suffice.

The system of 2D shallow water equations is obtained by integrating the Navier Stokes equations over depth and replacing the bed stress by a velocity squared resistance term in the two orthogonal directions. Various flood flow models can be constructed, depending on which terms in the governing equations are assumed to be negligible in comparison with the remaining terms. A 2D flow over inundated urban flood plain is assumed to be a slow, shallow phenomenon and therefore the convective acceleration terms can be assumed to be sufficiently small compared to the other terms so that they can be ignored.

The numerical solution of a given set of differential equations is defined by the discretisation strategy, the mesh used and the numerical scheme implemented. The shallow water equations system, having a hyperbolic character in time, represents an evolutionary problem in the form of propagating waves. Therefore a time marching procedure starting with a given initial condition in space, supplemented with boundary conditions along the time path is the proper mathematical conceptual treatment. The spatial discretization can be made with one of the following approaches: Finite Difference, Finite Volume and Finite Element methods.

A non-convective wave 2D model is developed in this research. The PDEs of the governing equations are transformed to difference equations on a regular Cartesian grid and a finite difference method is implemented for the numerical solution. A two point forward spatial and temporal difference scheme is adopted based on a uniform time step.

The governing equations are solved using the ADI procedure. The main characteristics of this procedure is that the calculation is split into two series of 'one-dimensional' calculations which are mutually orthogonal. The method provides greater superiority to the explicit finite difference method due to the high computational efficiency and thus less computing time. This efficiency is attained as the method involves a tridiagonal matrix and uses larger time steps due to its unconditional stability provided the finite difference scheme is properly formulated.

The water depth of a grid cell is calculated as the average depth over the whole cell. When the cell first receives water, the wetting front edge usually lies within the cell. In most cases, only part of the cell will be wetted at that time step. To avoid the problem of negative water depth and creation of isolated artificial wet patches, the wetting process is controlled by a wetting parameter which does not allow the water to flow out of the cell until the wetting front has crossed the cell.

Several numerical experiments on different hypothetical case studies show that the developed Non-convective wave 2D model provides stable results even for Courant numbers much greater than one. Although the numerical scheme used is believed to be stable for practical purposes, the time step is limited for accuracy purposes. The model uses adaptive time step for increasing model efficiency. It has the ability to halve or double the time step; halving in order to meet the convergence criterion, and doubling after a certain number of time steps without halving.

The modelling system is tested on different case studies. The results of the tests demonstrate that the developed model produced results which are in good agreement with solutions of empirical analytical equation for the wave front velocity (the empirical Chézy equation) and with the results of a reference model which implements the full shallow water equations (MIKE21). The tests show that, whilst neglecting convective acceleration terms which lead to local inaccuracies mainly for velocity predictions, the model is capable of predicting flood inundation depths successfully.

Coupling the 2D model with a 1D model
2D urban flood models have a sound physical basis in terms of the shallow-water equations. Though 1D flood modelling has been used as a standard industry practice for more than 30 years, it is generally known that 1D models cannot properly simulate the complex phenomenon of the interaction between the flows in the sewers and the flows above ground. This is because these models neglect some important aspects and suffer from a number of drawbacks when applied to floodplain flows, such as the inability to simulate lateral diffusion of the flood wave, the discretization of topography as cross-sections rather than as a surface and the subjectivity of cross-section location and orientation (Hunter et al. 2007; Kuiry 2010). Due to these reasons 1D-2D coupled model approaches are gaining greater attention; they can more adequately assess the performance of storm water or combined sewer systems and the associated surface flooding in order to predict the value of potential flood damages. This approach enables elements which are essentially one-dimensional (e.g. drains, culverts, channels) to be modelled explicitly, while overland processes are modelled with a two-dimensional schematization.

The developed 2D model is coupled with the Storm Water Management Model (SWMM5) developed by the United States Environmental Protection Agency (USEPA) (Rossman et al., 2005). In this way the complex nature of the interaction between surcharged sewer and flows associated with urban flooding can be simulated.

In the coupled model, the hydrological rainfall-runoff process and routing of flows in drainage pipes are performed using the 1D sewer network model. When the capacity of the pipe network is exceeded, excess flow spills into the two-dimensional model domain from the manholes and is then routed using the Non-convective wave 2D overland flow model. Both models use different numerical schemes and time steps with the discharge through the manholes forming the linkage of the models. The source code of SWMM5 is modified such that the surcharge in the sewer network is represented in terms of hydraulic head rather than overflow volumes. The interacting discharges are determined using weir or orifice equations by taking account of the hydraulic head at manholes and the aboveground water surface for every time step of the sewer network model such that the manhole discharges are treated as point sinks or sources in the 2D model within the same time interval.

The ability of the coupled model to simulate the interaction of below ground sewer network flow and above ground surface flow is studied within two case studies. The coupled model is used to simulate the drainage system of the Segunbagicha catchment and the associated floodplain of Dhaka, Bangladesh, and the drainage system and of flood plain of a catchment along Sukhumvit Road in the inner part of Bangkok, Thailand. The coupled models developed suffer from a lack of field data to calibrate the models and validate the results. Therefore, the results of the coupled models are not assessed through parameters such as flood area extent, location and flood depth, and velocity time series in comparisons with field observations. However, the results are compared to the results of previous studies using a coupled MOUSE and 2D MIKE21 model in the case of Dhaka and to limited measured water levels and eye witness flood depth observations in the case of Bangkok. From the two case studies, it can be concluded that the couple 1D-2D model has the capability to simulate the complex interaction between the sewer flow and the surcharge-induced inundation.

Modifiying the equation of the 2D surface flow model to improve flood model simulations in coarse grid models

The relationship between water surface elevation and volume of water that can be stored within a grid cell is a simple linear relationship when the average elevation is considered for a coarse grid in which the sub-grid topography variation within the grid is not considered. In such cases the relationship between water surface elevation and the flow area is also linear. However, these relationships do not represent the actual volume of storage nor the flow area and thus disregard their effect on the flood propagation. To improve flood forecasts in geometrically complex urban environment using coarse grid models, the non-linear relationships between volume and water depth and flow area and water depth should therefore be taken in to account. In an effort to incorporate these non-linear relationships the 2D flood model is modified. The continuity and momentum equations which describe the 2D surface flow model are rewritten in such a way that the volume and area are expressed as a non-simple function of water depth instead of a fixed plane area of the coarse grid. The original equations defined for flow at a point are reinterpreted in terms of a finite cell. The modified equations make use of the volume-depth and flow-area-depth relationships that can be extracted from fine resolution DEM. The model is developed so that these relationships can be extracted for desired coarse grid size from available fine resolution DEM.

The performance of the modified modelling approach for coarse grid models is tested on a case study. Two models, one with the modified approach taking volume-depth and flow-area-depth relationships in to account and another one with the standard linear approach, are used to simulate shallow inundation originating from an inflow to an area. The case study consists of a rectangle area with dense urban development and a topologically complex network of minor roads mixed with some steep sections of roads and local depressions where water may pond. A fine grid resolution model of 2 by 2 meters is considered as a benchmark model for comparison of coarse grid models. A 4 by 4 meters resolution coarse grid is used for the standard approach. For the modified approach three coarse grid models of 4 by 4 meters, 8 by 8 meters and 16 by 16 meters, are developed. Comparing the flood simulation results of the coarse grid models with the results of the benchmark model shows that the modified modelling approach enables the use of coarse grid models without significant loss of accuracy and with the advantage of reducing the computational time. Compared with the standard coarse model, the modified model approach predicted flood depths more accurately. The results also revealed that flood velocities are more sensitive to model grid resolution than the flood depths. The modified model approach for coarse grid modelling is a promising method; it enables the use of coarse grid models which considerable reduces the computational time and yet the influences of small scale urban features on the flow are taken in to account in terms of storage-depth and flow-area-depth relationships.

Infiltration process in the 2D surface flow model
With the emphasis given to flood peak control measures, incorporating surface water infiltration process into urban flood simulation models will help to evaluate their effect on urban flooding. To investigate how infiltration process affects urban flooding, an algorithm to simulate infiltration based on modified Horton method is incorporated in the 2D model. The algorithms requires input of initial and equilibrium infiltration capacities, the rate of decline of the infiltration capacity and initial soil water content appropriate to antecedent condition. The actual depth of infiltration accumulated during a time step is calculated based on the magnitude of infiltration at the beginning and end of the time step relative to the depth of water available in the time step. Three possible cases in relation to the available water depth per time step and potential infiltration rate are considered. These are (i) when the available depth of water per time step is greater than or equal to the

infiltration rate at the beginning of the time step, (ii) when the available depth of water per time step is less than the infiltration rate at the beginning of the time step but greater than the infiltration rate at the end of the time step and (iii) when the available depth of water per time step is less than or equal to the infiltration rate at the end of the time step.

The 2D modelling system with infiltration process is used to simulate rainfall-runoff processes on a flat plane and a plane with a slope of 2% in one direction of space. The cases simulated include different rainfall events and different initial infiltration capacities. The simulation results show that the algorithm based on modified Horton method allows for satisfactory simulation of the infiltration process. Expressing infiltration behaviour as a function of soil water content and the infiltration rate decays with time allows for recovery of infiltration capacity during intermittent rainfall. The result of the simulation also shows that surfaces with high initial infiltration capacity reduce the amount and delay the generation of surface runoff.

8.2 Recommendation

Urban flood models play an important role in urban flood risk management. Urban flood risk management requires an identification of the risk, the development of strategies to reduce that risk, and the creation of policies and programmes to put these strategies into effect. Urban flood models enable assessment of potential risk and understanding of the consequences of a flooding event. They enable a better understanding of the behaviour of urban drainage systems so that different strategies to mitigate structural and operational problems can be developed, tested and evaluated to assess their ability for reducing exposure to flood risk. There are many urban flood models developed in the past and existing urban flood models greatly vary depending on their purpose, the governing equations, the numerical scheme used and dimensionality (1D, 2D and 3D). Some models are developed for research purpose and even more models are developed as a commercial software package. This research focuses on development and application of an urban flood model at different topographical resolutions. An approach enabling simulation of urban floods using coarse grid models which retain information extracted from fine grid digital terrain model has been proposed and tested. The developed model and proposed approaches have several opportunities for further improvement and research. The following recommendations indicate some main aspects of urban flood modelling for further research:

1.   In literature the relative advantage and weakness of flood inundation models which implement simplified or reduced 2D shallow water equations, such as non-convective acceleration wave models and non-local

acceleration wave models, compared to models which solve the full shallow equations are given. These are drawn mostly from either hypothetical case studies or from models developed for real case studies which are not rigorously calibrated and validated. This is mainly due to the difficulty in obtaining a complete data set of a flood event for calibration and validation of flood models. In this research the urban flood model is developed assuming two-dimensional flow over inundated urban flood plain is a slow, shallow phenomenon in which the convective acceleration terms can be assumed to be small compared to the other terms in the momentum equations and therefore can be ignored. It was observed that in most application the developed non-convective acceleration wave model performed almost in the same way as models which solve full shallow water equations. It would be a great contribution to science if the relative weaknesses and strengths of such simplified 2D models can be verified compared to full equation models with experimental data sets. To be able to collect the required data set for this purpose, it is recommended to set up an experimental urban catchment.

2. One of the important parameters in urban flood modelling is the roughness coefficient. In 2D urban flood modelling roughness coefficients can be estimated for each grid based on land use types. This is straight forward if a grid contains a single land use type, though in coarse grid models a single grid often contains different land use types. In such cases usually the average of the roughness coefficients corresponding to the various land use types present in the coarse grid is taken to estimate the combined roughness coefficient. However, this may not represent the change in effective roughness across the coarse grid cell. Researching a better way of estimating effective roughness coefficient for combined land use types will greatly contribute towards accurate urban flood simulation.

3. The developed approach for generalization of grid to develop a coarse grid model enables representation of storage and flow-area within each coarse grid in the form of storage-depth and flow-area-depth relationships. However, in large coarse grids it is possible that these relationships may be flow direction dependent. For example, a ridge located in the middle of a coarse grid can block water so that it is only stored in part of the coarse grid which may necessarily not be the lowest part of the grid. Simple storage-depth and flow-area-depth relationships do not represent these

effects correctly. In such cases, more accurate relationships may be developed by taking in to account the flow direction in each computational grid. In other words, it is recommended to develop flow direction dependent storage-depth and flow-area-depth relationships and use the appropriate relationship on each time step during simulation. This obviously increases the computational expense and therefore it is recommended to evaluate the accuracy gained versus the increase computational expense.

4.   Another area to explore in relation to storage-depth and flow-area-depth relationships for coarse grid models concerns the accuracy of these relationships in topographies with steep slopes and insignificant variation in elevation of sub-grid cells. Computing the storage-depth relationships in the same way as for relatively flat topographies introduce inaccuracies. This is due to the fact that computation of storage-depth relationships assumes the flood water start accumulating from the lowest part of the coarse grid and gradually covers the grid as the depth of flood water increases, whereas the flood water may be stored across the sloping plane with uniform depth. If the sub-grid cells contained in the coarse grid do not vary significantly in elevation, one way forward to improve the accuracy of the storage-depth and the flow-area-depth relationships is recommended as follows. Firstly computing the average slope of the coarse grid in both directions (the orthogonal directions) and secondly projecting the sloping plane (with the average slope computed) to a horizontal plane and adjust the elevations of the sub-grid cells according to the projection. Finally, flow direction dependant storage-depth and flow-area-depth relationship can be computed in the same way as flat topographies.

5.   The approach for generalization of grid to develop a coarse grid model also offers the incorporation of effects caused by structures which allow flood to pass through such as bridges, culverts and tunnels. These effects cannot be readily represented by DTM unless information is available on location, size and elevation of such structures. It is recommended to test the applicability of the approach in such cases.

6.   In developing the coupled 1D-2D model, the crest elevation of the connecting sewer node (manhole) is assumed to be equal to the elevation of the grid cell where the node is located. This is done so in order to

minimize inaccuracies in calculating the interacting discharge due to inconsistencies between the mean elevation of the grid cell and ground level of the sewer node. This cannot be done in a straight forward manner in the case of coupling coarse grid models with 1D models. Firstly because the approach developed in this research considers volume-depth relationship instead of mean elevation of the grid. Secondly, in coarse grids more than one connecting sewer nodes with different ground elevations may exist. It is therefore recommended to further develop the coupling method so that it is applicable to coarse grid models as well.

References

Abbott, M. (1980), *Computational hydraulics. Elements of the theory of free surface flows*, Pitman Publishing Ltd., London.

Abbott, M., and Ionescu, F. (1967), On the numerical computation of nearly horizontal flows, *Journal of Hydraulic Research, 5*(2), 97-117.

Abdullah, A., Vojinovic, Z., Price, R., and Aziz, N. (2011), A methodology for processing raw LiDAR data to support urban flood modelling framework, *Journal of Hydroinformatics 14*(1), 75 - 92.

Ahmad, S., and Simonovic, S. P. (2006), An intelligent decision support system for management of floods, *Water resources management, 20*(3), 391-410.

Ahmed, F. (2008), Urban Flood, its Effects and Management Options: A Case Study of Dhaka city, UNESCO-IHE Institute for Water Education, Delft.

Akan, A. O. (1993), *Urban stormwater hydrology: a guide to engineering calculations*, CRC.

Alam, M., and Rabbani, M. (2007), Vulnerabilities and responses to climate change for Dhaka, *Environment and Urbanization, 19*(1), 81.

Alcrudo, F. (2004), Mathematical modelling techniques for flood propagation in urban areas, IMPACT Project technical report (available at http://www.impact-project.net/AnnexII_DetailedTechnicalReports/AnnexII_PartB_WP3/Modelling_t echniques_for_urban_flooding.pdf).

Alcrudo, F., Garcia-Navarro, P., and Saviron, J. M. (1992), Flux difference splitting for 1D open channel flow equations, *International Journal for Numerical Methods in Fluids, 14*(9), 1009-1018.

Aldrighetti, E. (2007), Computational hydraulic techniques for the Saint Venant Equations in arbitrarily shaped geometry, Universita degli Studi di Trento, Trento.

Andjelkovic, I. (2001), *Guidelines on non-structural measures in urban flood management*, International Hydrological Programme -V Technical documents in hydrology, No 50, UNESCO, Paris.

Aronica, G., and Lanza, L. (2005), Drainage efficiency in urban areas: a case study, *Hydrological Processes, 19*(5), 1105-1119.

Balmforth, D., Digman, C., Butler, D., and Shaffer, P. (2006), Defra Integrated Urban Drainage Pilots, Scoping Study, edited, Defra Publications.

References

Barredo, J. I., Lavalle, C., Sagris, V., and Engelen, G. (2005), Representing future urban and regional scenarios for flood hazard mitigation, paper presented at 45th Congress of the European Regional Science Association, Land Use and Water Management in a Sustainable Network Society, Vrije Universiteit, Amsterdam, 23-27 August 2005. .

Bauer, S., W (1974), A modified Horton equation for infiltration during intermittent rainfall, *Hydrological Sciences Journal, 19*(2), 219-225.

Bolle, A., Demuynck, A., Bouteligier, R., Bosch, S., Verwey, A., and Berlamont, J. (2006), Hydraulic Modelling of the Two-directional Interaction between Sewer and River Systems, paper presented at 7th International Conference on Urban Drainage Modelling, Monash University, Melbourne, Austrlia.

Burrough, P. A., McDonnell, R. A., and McDonnell, R. (1998), *Principles of geographical information systems*, Oxford university press Oxford.

Campana, N. A., and Tucci, C. E. M. (2001), Predicting floods from urban development scenarios: case study of the Dilúvio Basin, Porto Alegre, Brazil, *Urban Water, 3*(1), 113-124.

Carr, R. S., and Smith, G. P. (2006), Linking of 2D and Pipe hydraulic models at fine spatial scales, paper presented at 7th International Conference on Urban Drainage Modelling Monash University, Melbourne, Austrlia.

Casulli, V. (1990), Semi-implicit finite difference methods for the two-dimensional shallow water equations, *Journal of Computational Physics, 86*(1), 56-74.

Chanson, H. (2004), *Environmental hydraulics of open channel flows*, Butterworth-Heinemann.

Chaudhry, M. H. (1987), Applied hydraulic transients, *Second edition. Van Nostrand Reinhold Co. New York. 1987. 521.*

Chaudhry, M. H. (1993), Open-channel flow, *Pretice Hall, New Jersey.*

Chen, A., Hsu, M., Chen, T., and Chang, T. (2005), An integrated inundation model for highly developed urban areas, *Water Science and Technology, 51*(2), 221-230.

Chen, A., Djordjevi , S., Leandro, J., and Savi , D. (2007), The urban inundation model with bidirectional flow interaction between 2D overland surface and 1D sewer networks, *NOVATECH 2007*, 465-472.

Chen, A., Evans, B., Djordjević, S., and Savić, D. (2012), Multi-layered coarse grid modelling in 2D urban flood simulations, *Journal of Hydrology.*

Chocat, B., Krebs, P., Marsalek, J., Rauch, W., and Schilling, W. (2001), Urban drainage redefined: from stormwater removal to integrated management, *Water science and technology: a journal of the International Association on Water Pollution Research, 43*(5), 61.

Crossely, A. (1999), Accurate and efficient numerical solutions for the Saint Venant equations of open channel flow, PhD thesis, University of Nottingham, Nottingam.

Cunge, J., Holly, F., and Verwey, A. (1980), *Practical aspects of computational river hydraulics*, Pitman Publishing Ltd., London.

Dabberdt, W. F., Hales, J., Zubrick, S., Crook, A., Krajewski, W., Doran, J. C., Mueller, C., King, C., Keener, R. N., and Bornstein, R. (2000), Forecast issues in the urban zone: Report of the 10th Prospectus Development Team of the US Weather Research Program, *BULLETIN-AMERICAN METEOROLOGICAL SOCIETY*, *81*(9), 2047-2064.

De Sherbinin, A., Schiller, A., and Pulsipher, A. (2007), The vulnerability of global cities to climate hazards, *Environment and Urbanization*, *19*(1), 39-64.

DHI Group (2009a), *MOUSE pipe flow reference manual*, DHI Software, Horsolm, Debmark.

DHI Group (2009b), *MIKE 21 FLOW MODEL: Hydrodynamic Module Scientific Documentation*, DHI Software, Horsolm, Debmark.

Djordjević, S., Prodanovic, D., Maksimovic, C., Ivetic, M., and Savić, D. (2005), SIPSON: Simulation of interaction between pipe flow and surface overland flow in networks, *Water Science and Technology*, 275-283.

Douglas, I., Alam, K., Maghenda, M. A., Mcdonnell, Y., McLean, L., and Campbell, J. (2008), Unjust waters: climate change, flooding and the urban poor in Africa, *Environment and Urbanization*, *20*(1), 187-205.

Dubrovin, T., Keskisarja, V., Sane, M., and Silander, J. (2006), Flood Management in Finland–introduction of a new information system, in *7th International Conference on Hydroinformatics, Nice, France*, edited.

Esteves, M., Faucher, X., Galle, S., and Vauclin, M. (2000), Overland flow and infiltration modelling for small plots during unsteady rain: numerical results versus observed values, *Journal of Hydrology*, *228*(3-4), 265-282.

Evans, B. (2010), A Multilayered Approach to Two-Dimensional Urban Flood Modelling, University of Exeter.

Falconer, R. A. (1980), Numerical modeling of tidal circulation in harbors, *Journal of the Waterway Port Coastal and Ocean Division*, *106*(1), 31-48.

Ferziger, J. H., and Peri , M. (1999), *Computational methods for fluid dynamics*, Springer Berlin.

References

Few, R., Ahern, M., Matthies, F., and Kovats, S. (2004), *Floods, health and climate change: a strategic review*, Tyndall Centre for Climate Change Research Norwich.

Garcia-Navarro, P., and Brufau, P. (2006), Numerical methods for the shallow water equations: 2D approach, in *River basin modeling for flood risk mitigation*, edited by Knight, D. W. and Shamseldin, A. Y., pp. 409-428, Taylor & Francis.

Garcia, R., and Kahawita, R. A. (1986), Numerical solution of the St. Venant equations with the MacCormack finite-difference scheme, *International Journal for Numerical Methods in Fluids*, 6(5), 259-274.

Glaister, P. (1988), Approximate Riemann solutions of the shallow water equations, *Journal of Hydraulic Research*, 26(3), 293-306.

Green, W. H., and Ampt, G. (1911), Studies on soil physics, *J. Agric. Sci*, 4(1), 1-24.

Grum, M., Jørgensen, A. T., Johansen, R. M., and Linde, J. J. (2006), The effect of climate change on urban drainage: an evaluation based on regional climate model simulations, *Water Science and Technology*, 54(6-7), 9-15.

Guinot, V., and Soares-Frazão, S. (2006), Flux and source term discretization in two-dimensional shallow water models with porosity on unstructured grids, *International Journal for Numerical Methods in Fluids*, 50(3), 309-345.

Hervouet, J.-M. (2007), *Hydrodynamics of Free Surface Flows - modelling with the finite element method*, John Wiley & Sons Ltd.

Hervouet, J.-M., Samie, R., and Moreau, B. (2000), Modelling urban areas in dam-break flood-wave numerical simulations, paper presented at International Seminar and Workshop on Rescue Actions Based on Dambreak Flow Anlysis, Sein^ajoki, Finland, 1–6 October, 2000.

Hicks, F., and Peacock, T. (2005), Suitability of HEC-RAS for flood forecasting, *Canadian Water Resources Journal*, 30(2), 159-174.

Horritt, M., and Bates, P. (2002), Evaluation of 1D and 2D numerical models for predicting river flood inundation, *Journal of Hydrology*, 268(1-4), 87-99.

Horton, R. E. (1939), Analysis of runoff plot experiments with varying infiltration capacity, *Trans. Am. Geophys. Union*, 20, 693-711.

Hromadka II, T., McCuen, R., and Yen, C. (1987), Comparison of overland flow hydrograph models, *Journal of Hydraulic Engineering*, 113, 1422.

Hsu, M., Chen, S., and Chang, T. (2000), Inundation simulation for urban drainage basin with storm sewer system, *Journal of Hydrology*, 234(1-2), 21-37.

Huber, W. C., Dickinson, R. E., and Barnwell Jr, T. O. (1988), Storm water management model; version 4, *Environmental Protection Agency, United States*.

Hunter, N., Bates, P., Horritt, M., and Wilson, M. (2007), Simple spatially-distributed models for predicting flood inundation: A review, *Geomorphology*, *90*(3-4), 208-225.

Hunter, N., et al. (2008), Benchmarking 2D hydraulic models for urban flooding, *Proceedings of the Institution of Civil Engineers*, *161*(1), 13-30.

Jain, M. K., and Singh, V. P. (2005), DEM-based modelling of surface runoff using diffusion wave equation, *Journal of Hydrology*, *302*(1-4), 107-126.

Jha, A., Bloch, R., and Lamond, J. (Eds.) (2012), *Cities and Flooding: A Guide to Integrated Urban Flood Risk Management for the 21st Century*, The World Bank.

Jha, A., Lamond, J., Bloch, R., Bhattacharya, N., Lopez, A., Papachristodoulou, N., Bird, A., Proverbs, D., Davies, J., and Barker, R. (2011), *Five feet high and rising: cities and flooding in the 21st century*, World Bank.

Kolsky, P., and Butler, D. (2002), Performance indicators for urban storm drainage in developing countries, *Urban Water*, *4*(2), 137-144.

Kuiry, S. N., Sen, D., and Bates, P. D. (2010), Coupled 1D–Quasi-2D Flood Inundation Model with Unstructured Grids, *Journal of Hydraulic Engineering*, *136*, 493.

Kutija, V. (1996), Flow adaptive schemes, Doctoral thesis, Delft University of Technology and International Institute for Infrastructural, Hydraulic and Environmental Engineering, Delft, The Netherlands.

Leandro, J. (2008), Advanced Modelling of Flooding in Urban Areas Integrated 1D/1D and 1D/2D Models, PhD thesis thesis, University of Exeter, Exeter.

Leandro, J., Chen, A. S., DjordjeviÄ, S., and SaviÄ, D. A. (2009), Comparison of 1D/1D and 1D/2D coupled (sewer/surface) hydraulic models for urban flood simulation, *Journal of Hydraulic Engineering*, *135*(6), 495-504.

Leandro, J., Djordjević, S., Chen, A. S., and Savić, D. A. (2009), Comparison of 1D/1D and 1D/2D Coupled (Sewer/Surface) Hydraulic Models for Urban Flood Simulation, *Journal of Hydraulic Engineering - ASCE*, *136*(6), 495-504.

Leendertse, J. (1967), Aspects of a computational model for well mixed estuaries and coastal seas, *RM-5924-Pr, The Rand Corp.*

Lhomme, J., Bouvier, C., Mignot, E., and Paquier, A. (2006), One-dimensional GIS-based model compared with a two-dimensional model in urban floods simulation, *Water Science and Technology*, *54*(6-7), 83-91.

Liggett, J. A., and Cunge, J. A. (1975), Numerical methods of solution of the unsteady flow equations, *Unsteady flow in open channels*, *1*, 89-178.

Lin, B., Wicks, J., Falconer, R., and Adams, K. (2006), Integrating 1D and 2D hydrodynamic models for flood simulation.

MacCormack, R. W. (1969), The effect of viscosity in hypervelocity impact cratering, *Frontiers of Computational Fluid Dynamics*, 27-44.

Mark, O., Weesakul, S., Apirumanekul, C., Aroonnet, S. B., and Djordjević, S. (2004), Potential and limitations of 1D modelling of urban flooding, *Journal of Hydrology, 299*(3-4), 284-299.

Meselhe, E., and Holly Jr, F. (1997), Invalidity of Preissmann scheme for transcritical flow, *Journal of Hydraulic Engineering, 123*, 652.

Messner, F., Penning-Rowsell, E., Green, C., Meyer, V., Tunstall, S., and Van der Veen, A. (2007), Evaluating flood damages: guidance and recommendations on principles and methods, *FLOODsite-Report T, 9*.

Microsoft Visual Studio 2008: Visual C++, Microsoft. Accessed on 28, January, 2013, from http://msdn.microsoft.com/en-us/library/60k1461a(v=vs.90).aspx

Mignot, E., Paquier, A., and Haider, S. (2006), Modeling floods in a dense urban area using 2D shallow water equations, *Journal of Hydrology, 327*(1), 186-199.

Neal, J. C., Bates, P. D., Fewtrell, T. J., Hunter, N. M., Wilson, M. D., and Horritt, M. S. (2009), Distributed whole city water level measurements from the Carlisle 2005 urban flood event and comparison with hydraulic model simulations, *Journal of Hydrology, 368*(1-4), 42-55.

Néelz, S., and Pender, G. (2008), Grid resolution dependency in inundation modelling: A case study, in *Flood Risk Management: Research and Practice*, edited by William, A., Paul, S., Jackie, H. and Stephen, H., p. 29, CRC Press.

Néelz, S., and Pender, G. (2010), Benchmarking of 2D Hydraulic Modelling Packages*Science Report SC080035/R2*, 169 pp, Environment Agency, Bristol, UK.

Nikolov, N., Minkov, I., Dimitrov, D., Mincheva, S., and Mirchev, M. (1978), Hydraulic calculation of a submerged broad-crested weir, *Power Technology and Engineering (formerly Hydrotechnical Construction), 12*(6), 631-634.

O'Loughlin, G., and Seneviratne, I. (2006), Lessons from Modelling Piped Urban Drainage Catchments In : Urban Drainage Modelling and Water Sensitive Urban Design, Melbourne.

Parliamentary Office of Science and Technology (Parliamentary Office of Science and Technology) (2007), Urban Flooding, *Postnote Number 289*, Parliamentary Office of Science and Technology, London, (available at www.parliament.uk/documents/upload/postpn289.pdf).

Peaceman, D., and Rachford, H. J. (1955), The numerical solution of parabolic and elliptic differential equations, *Journal of the Society for Industrial and Applied Mathematics*, *3*(1), 28-41.

Phillips, B., Yu, S., Thompson, G., and de Silva, N. (2005), 1D and 2D Modelling of Urban Drainage Systems using XP-SWMM and TUFLOW, paper presented at 10th International conference on urban drainage, Copenhagen, Denmark.

Ponce, V. (1990), Generalized diffusion wave equation with inertial effects, *Water Resources Research*, *26*(5), 1099-1101.

Potter, M., and Wiggert, D. (2002), Mechanics of Fluids Brooks, *Cole, Pacific Grove, CA, USA.*

Price, R. K. (2009a), Volume-conservative nonlinear flood routing, *Journal of Hydraulic Engineering*, *135*, 838.

Price, R. K. (2009b), An optimized routing model for flood forecasting, *Water Resources Research*, *45*(2), W02426.

Price, R. K., and Vojinovic, Z. (2010), *Urban Hydroinformatics: Data, Models and Decision Support for Integrated Urban Water Management*, International Water Assn.

Ramos, H., and Almeida, A. B. (1987), Backwater Effects with a Diffusion Type Modeling, paper presented at Proc. IAHR Seminar on Wave Analysis and Generation in Laboratory Basins, 22nd IAHR Congress, Lausanne, Switzerland.

Rossman, L. A. (2006), Storm Water Management Model, Quality Assurance Report: Dynamic Wave Flow Routing, National Risk Management Research Laboratory, Office Of Research And Development, U.S. Environmental Protection Agency, Cincinnati. Retrieved from http://www.epa.gov/nrmrl/wswrd/wq/models/swmm/#Downloads

Rossman, L. A., Dickinson, R. E., Schade, T., Chan, C., Burgess, E. H., and Huber, W. C. (2005), SWMM 5: The USEPA's Newest Tool for Urban Drainage Analysis, in *Proceedings, 10th International Conference on Urban Drainage*, edited, Copenhagen, Technical University of Denmark.

Sart, C., Baume, J. P., Malaterre, P. O., and Guinot, V. (2010), Adaptation of Preissmann's scheme for transcritical open channel flows, *Journal of Hydraulic Research*, *48*(4), 428-440.

Schmitt, T. G., Thomas, M., and Ettrich, N. (2004), Analysis and modeling of flooding in urban drainage systems, *Journal of Hydrology*, *299*(3-4), 300-311.

Smith, G. D. (1985), *Numerical solution of partial differential equations: finite difference methods*, Oxford University Press, USA.

References

Spry, R. B., and Zhang, S. (2006), Modelling of Drainage Systems and Overland Flowpaths at Catchment Scales, paper presented at 7th International Conference on Urban Drainage Modelling Monash University, Melbourne, Austrlia.

Stelling, G. S., Wiersma, A., and Willemse, J. (1986), Practical aspects of accurate tidal computations, *Journal of Hydraulic Engineering, 112*, 802.

Tayfur, G., Kavvas, M. L., Govindaraju, R. S., and Storm, D. E. (1993), Applicability of St. Venant Equations for Two-Dimensional Overland Flows over Rough Infiltrating Surfaces, *Journal of Hydraulic Engineering, 119*, 51.

Unidata Program Center Network Common Data Form, Unidata Program Center, Boulder, Colorado. Accessed on 28, January, 2013, from http://www.unidata.ucar.edu/software/netcdf/

US Army Corps of Engineers (1993), Engineering and Ddesign: River Hydraulics, in *Engineering Manual*, edited, U. S. Army Corps of Engineers, Washington, DC.

Vaes, G., Bouteligier, R., Herbos, P., and Berlamont, J. (2004), Modelling of floods caused by urban drainage systems, paper presented at 6th int. conf. on Urban Drainage Modelling, Dresden, Germany.

Versteeg, H. K., and Malalasekera, W. (2007), *An introduction to computational fluid dynamics: the finite volume method*, Prentice Hall.

Verwey, A., Muttil, N., Liong, S., and He, S. (2008), Implementing an Urban Rainfall-runoff Concept in SOBEK for a Catchment in Singapore, *Proceedings of Water Down Under 2008*, 36.

Villemonte, J. R. (1947), Submerged-weir discharge studies, *Engineering News-Record*, 866-869.

Vojinovic, Z., and van Teeffelen, J. (2007), An integrated stormwater management approach for small islands in tropical climates, *Urban Water Journal, 4*(3), 211-231.

Vojinovic, Z., and Tutulic, D. (2009), On the use of 1D and coupled 1D-2D modelling approaches for assessment of flood damage in urban areas, *Urban Water Journal, 6*(3), 183-199.

Vojinovic, Z., Ediriweera, J., and Fikri, A. (2008), An approach to the model-based spatial assessment of damages caused by urban floods, paper presented at 11th International Conference on Urban Drainage, Edinburgh, Scotland, UK.

Vojinovic, Z., Bonillo, J., Kaushik, C., and Price, R. (2006), Modelling flow transitions at street junctions with 1D and 2D models. , paper presented at 7th International Conference on Hydroinformatics, Nice, FRANCE

Vojinovic, Z., Seyoum, S., Mwalwaka, J., and Price, R. (2011), Effects of model schematisation, geometry and parameter values on urban flood modelling, *Water Science and Technology*, *63*(3), 462-467.

Vreugdenhil, C. B. (1994), *Numerical methods for shallow-water flow*, Springer.

Wallingford-Software now MWHSoft (2004), Does traditional calibration hide errors in your demand analysis?, *news Articles (available at http://www.innovyze.com/news/fullarticle.aspx?id=260)*.

Watson, R., Zinyowera, M., and Moss, R. (1998), *The regional impacts of climate change: an assessment of vulnerability*, Cambridge Univ Pr.

Wechsler, S. (2007), Uncertainties associated with digital elevation models for hydrologic applications: a review, *Hydrology and Earth System Sciences*, *11*(4), 1481-1500.

Wood, J. (1996), The geomorphological characterisation of digital elevation models, *Advances in Cancer Research, 104*.

Wright, N., Villanueva, I., Bates, P., Mason, D., Wilson, M., Pender, G., and Neelz, S. (2008), Case Study of the Use of Remotely Sensed Data for Modeling Flood Inundation on the River Severn, UK, *Journal of Hydraulic Engineering, 134*, 533.

Yu, D., and Lane, S. (2006 (a)), Urban fluvial flood modelling using a two dimensional diffusion wave treatment, part 1: mesh resolution effects, *Hydrological Processes*, *20*(7), 1541-1565.

Yu, D., and Lane, S. (2006 (b)), Urban fluvial flood modelling using a two-dimensional diffusion-wave treatment, part 2: development of a sub-grid-scale treatment, *Hydrological Processes, 20*(7), 1567-1583.

Zhang, W., and Cundy, T. W. (1989), Modeling of two-dimensional overland flow, *Water Resources Research*, *25*(9), 2019-2035.

Zhilin, L. (2008), Multi-Scale Digital Terrain Modelling and Analysis, *Advances in Digital Terrain Analysis*, 59-83.

Table of Figures

List of Tables

About the Author

Solomon Dagnachew Seyoum is currently a lecturer at the department of Environmental Engineering and Water Technology of UNESCO-IHE Institute for Water Education with more than 15 years of research and industrial experience. He has a combined background in sanitary engineering, water resources engineering and hydroinformatics. Prior to joining UNESCO-IHE Solomon worked for government and consultancy firms in Ethiopia for which he was involved in the study, design, modelling and management of urban drainage, water supply, stormwater and wastewater systems. His role ranged from resident engineer to senior design engineer and team leader within various water resources development projects. Solomon completed his MSc in hydroinformatics in 2005 and after two years of practical experience in his country Ethiopia, he came back to pursue his PhD research. At UNESCO-IHE Solomon teaches in urban drainage and sewerage, design and modelling of urban drainage systems, numerical methods and related courses. In addition he supervises Master students and coordinates an online course on urban drainage and sewerage. Solomon's research interests include urban flood modelling, optimization of urban drainage and water supply distribution networks, development and application of advanced tools for simulation, design and management of water systems and urban water infrastructure asset management.

List of Publications

Journals papers

- Seyoum S.D., Vojinovic Z., Price R.K., and Weesakul S., (2012), "A Coupled 1D and Non-Inertia 2D Flood Inundation Model for Simulation of Urban Flooding", ASCE *Journal of Hydraulic Engineering* 138(1), 23-34.

- Vojinovic, Z., Seyoum S., Mawalwaka M., Fikri A., (2011), "Effect of Model Schematization, Geometry and Paraameter Values on Urban flood Modelling", Water Science and Technology 63(3), 462-467.

- Vojinovic, Z., Seyoum, S., Salum, M.H., Price, R.K., Fikri, A.F. and Abebe, Y., (2012)," Modelling urban floodplain inundation with different spatial resolution of 2D models", Journal of Hydroinformatics, IWA Publishing, doi:10.2166/hydro.2012.181

- Seyoum S.D., Vojinovic Z., and Price R.K., (2013), "Generalization of Topographical Resolutions for Two-Dimensional Urban Flood Modelling", ASCE Journal of Hydraulic Engineering, to be submitted.

- Kankanamalage R. K., Seyoum S.D., and Garcia H., (2013), "Design and Modelling of Simplified Sewers", Journal to be identified, to be submitted.

- Seyoum S.D., Price R.K., and Vojinovic Z., (2013), "Incorporating rainfall and infiltration process in flood modelling", Journal to be identified, planned.

- Sahilu S., Seyoum S.D., and Vojinovic Z., (2013), "Robust optimization of urban drainage systems under parameter uncertainties", to be submitted.

Conference proceedings

- Vojinovic Z., Seyoum S.D., Price R., Salum M., Fikri A. and Abebe Y., (2012), Terrain data processing and treatment of different features for urban flood modelling, Hydroinformatics conference, 2012, July14-18 , Hamburg , Germany

- Vojinovic Z., Anvarifar F., Matungulu H., Torres A.S., Seyoum S.D., Barreto W., Price R., (2012) Multi-criteria risk-based optimisation of urban drainage assets, Hydroinformatics conference, 2012, July14-18 , Hamburg , Germany

- Andino O., Pathirana A., Seyoum S.D. and Brdjanovic D., (2012) "Development and application of an optimization tool for urban drainage network design under uncertainty", 9th International Conference on Urban Drainage Modelling, Belgrade, Serbia, September 3-7.

- Hodzic A., Vojinovic Z., Seyoum S.D., Pathirana A., Meijer S.C.F., and Brdjanovic D., (2011), "Model-based evaluation of the urban wastewater infrastructure reconstruction options in a developing country: Case study Sarajevo in Bosnia and Herzegovina". 2nd IWA Development Congress, Kuala Lumpur, Malaysia.

- Seyoum S., Vojinovic Z. and Price, R., 2010, Urban Pluvial Flood Modelling: Development and Application, 9th International Conference on Hydroinformatics 2010, Tianjin, China September 7-11.

- Vojinovic Z., Matingulu H., Sanchez A., Seyoum S.D., and Barreto W., (2010), Multi-object Optimization of Urban Drainage Rehabilitation Measures using Evolutionary Algorithms, 9th International Conference on Hydroinformatics 2010, Tianjin, China September 7-11.

- Vojinovic, Z., Seyoum S., Salum M., Mawalwaka M., Price R.K., and Fikri A., (2010), Modelling Urban Floodplain Inundation with Different Spatial Resolution and Model Parameterisation, 9th International Conference on Hydroinformatics 2010, Tianjin, China September 7-11.

- Vojinovic, Z., Seyoum S., Mawalwaka M., Fikri A., (2009), Effect of Model Schematization, Geometry and Parameter Values on Urban flood Modelling, 8th International Conference on Urban Drainage Modelling, Tokyo, Japan September 7-12

- Seyoum S., Vojinovic Z. and Price, R., (2009), Integrated Modelling of Urban Wastewater Systems, 8th International Conference on Urban Drainage Modelling, Tokyo, Japan September 7-12

- Seyoum S. and Vojinovic Z., (2008), Integrated Urban Water Systems Modelling with a Simplified Surrogate Modular Approach, 11th International Conference on Urban Drainage, Edinburgh, Scotland, August 31-Septemeber 5.

Samenvatting

FRAMEWORK FOR DYNAMIC MODELLING OF URBAN FLOODS AT
DIFFERENT TOPOGRAPHICAL RESOLUTIONS

Solomon Dagnachew Seyoum

Overstromingen zijn een van de meest voorkomende en duurste natuurrampen in termen van menselijk lijden en economische schade. De gevolgen van overstromingen zijn vooral groot in dichtbevolkte stedelijke gebieden waar belangrijke infrastructuur aanwezig is. Naar verwachting zal het risico op en de gevolgen van overstroming in stedelijke gebieden toe nemen als gevolg van verdere verstedelijking van overstromingsgevoelige uiterwaarden. Bovendien wordt verwacht dat de regenval ten gevolge van klimaatverandering toeneemt, terwijl de afvoer capaciteit van verouderde rioleringen in bestaande stedelijke gebieden verder afneemt. De toegenomen kans op en consequenties van overstromingen vraagt om de ontwikkeling van kosteneffectieve maatregelen om overstromingen te voorkomen als onderdeel van plannen voor duurzaam stedelijke waterbeheer. Efficiënte voorspelling van mogelijke overstromingen in stedelijke gebieden is cruciaal voor de ontwikkeling van tegenmaatregelen. Computermodellen voor stedelijk waterbeheer proberen kwantitatief de eigenschappen en verspreiding van overstromingen te beschrijven die optreden als een grote hoeveelheden water door afwateringssystemen stromen in stedelijke gebieden.

Hoewel topografische data met hoge resolutie essentieel zijn voor gedetailleerde voorspellingen van overstromingen in stedelijke gebieden, wordt de data vaak teruggebracht tot lagere resoluties vanwege de lange rekentijdtijd nodig voor hoge resolutie berekeningen. Overstromingsmodellen hebben daarom vaak grovere resoluties en maken gebruik van gegeneraliseerde topografische data met een lagere resolutie. Echter, de generalisatie van topografische data kan leiden tot significante veranderingen in de topografie van het stedelijk landschap doordat dominante karakteristieken zoals gebouwen, muren en begroeiing uitgesmeerd of weggelaten worden. Als gevolg hiervan zullen computermodellen met een lage resolutie vaker simulaties van overstroming onnauwkeurige berekenen dan modellen met een hogere resolutie. Verschillende methoden zijn bedacht om de informatie die verkregen kan worden uit topografische data met een hoge resolutie te behouden in modellen met een grovere resolutie. Dit onderzoek richt zich op de ontwikkeling en toepassing van een methode om kleinschalige stedelijke karakteristieken te kunnen meenemen in lage resolutie twee dimensionale (2D) stedelijk overstromingsmodellen met als doel de nauwkeurigheid van de voorspellingen van overstromingen in geometrisch complexe gebieden te vergroten.

Voor dit onderzoek is software ontwikkeld voor de 2D oppervlaktestroming gebaseerd op de 2D golf vergelijkingen met verwaarlozing van de traagheidstermen. De software geeft de stedelijke topografie weer door middel van terreinhoogte in het midden en op de randen van de cellen van een rechthoekig Cartesisch rooster. Het waterniveau wordt bepaald in het midden van de cel en de afvoer (uitgedrukt in snelheden) op de celranden. Een impliciete eindige-differentie methode met alternerende richtingen wordt gebruikt voor het oplossen van de vergelijkingen. De software is getest in verschillende case studies en de resultaten laten zien dat het model goed overeenkomt een referentiemodel dat gebruikt maakt van de volledige ondiep water vergelijkingen (in dit geval MIKE21). De testen tonen aan dat, ondanks verwaarlozing van de convectieve versnellingstermen die leiden tot lokale onnauwkeurigheden in stroomsnelheden, het model in staat is om de waterdiepten goed te voorspellen.

Als onderdeel van dit onderzoek is een 1D-2D gekoppeld software systeem ontwikkeld voor het koppelen van het ontwikkelde overstromingsmodel aan het 'Storm Water Management Model' (SWMM5). De mogelijkheden van het gekoppelde software systeem zijn in twee case studies getest. Er zijn gekoppelde computermodellen ontwikkeld voor (i) het rioleringstelsel in het Segunbagicha stroomgebied en bijbehorende uiterwaarden in Dhaka, Bangladesh, en (ii) het rioleringsstelsel en uiterwaarden langs de Sukhumvit straat in het centrum van Bangkok, Thailand. Ondanks dat er niet voldoende metingen waren om de modellen uitgebreid te kalibreren en de resultaten te valideren, kan er op basis van eerdere studies en observaties geconcludeerd worden dat de modellen geschikt zijn voor het simuleren van overstromingen ten gevolge van de overdruk in rioleringssystemen.

Om overstromingen in complexe stedelijke gebieden beter te kunnen voorspellen moeten de modellen met grove rasters de niet-lineaire relaties tussen het volume en waterdiepte en tussen stroomoppervlak en waterdiepte meenemen. Om deze niet-lineaire relaties mee te nemen is de software van het 2D overstromingsmodel aangepast. De continuïteits- en impulsvergelijkingen in de software van het 2D oppervlaktestroom model zijn zodanig aangepast dat het volume en het stroomoppervlak meervoudige functies zijn van de waterdiepte en niet van een vast oppervlak in het grove raster. De aangepaste vergelijkingen maken gebruik van de relaties volume-waterdiepte en stroomoppervlakte-waterdiepte verkregen uit digitale hoogtemodellen met fijne resoluties. De software is in staat om deze relaties voor de gewenste grootte van het grove raster te verkrijgen vanuit het beschikbare digitale hoogtemodel met fijne resolutie.

Het gedrag van de aangepaste software voor de grove raster modellen is getest aan de hand van een case studie. Twee modellen, een met een aangepaste software om de niet-lineaire volume-diepte en stromingsoppervlakte-diepte relaties mee te nemen en de andere met een standaard lineaire benadering, zijn gebruikt om ondiepe overstromingen te simuleren door een nauwe instroomopening. De resultaten laten zien dat de aangepaste software het mogelijk maakt om grove raster modellen te gebruiken zonder veel verlies aan nauwkeurigheid, terwijl de rekentijd aanzienlijk wordt verkort.

Vanwege de nadruk die gegeven wordt aan maatregelen voor het beheersen van de piekafvoer door middel van het vergroten van de infiltratiecapaciteit, is het nodig om infiltratieprocessen van oppervlaktewater in de modellen op te nemen teneinde het effect van deze maatregelen op overstromingen in stedelijke gebieden te kunnen beoordelen. Hiertoe is een algoritme gebaseerd op een aangepaste Horton methode opgenomen in de software. De resultaten van de case studies waarin de neerslag-afvoer processen werden gesimuleerd laten zien dat het algoritme gebaseerd op de aangepaste Horton methode toereikend is voor het simuleren van het infiltratie proces.

Dit onderzoek was gericht op het ontwikkelen van een methode voor het verbeteren van simulaties met verschillende topografische resoluties, en draagt op de volgende manieren bij aan het onderzoeksveld voor het modeleren van overstromingen in stedelijke gebieden:

1. Het ontwikkelen en testen van nieuw software voor overstromingsmodellen in stedelijke gebieden die de 2D niet-convectieve golfvergelijkingen oplost voor geleidelijk variërende vrij oppervlaktestroming. De software heeft innovatieve eigenschappen zoals het gebruik van dubbelslagiteraties voor het bereiken van nauwkeurige oplossingen, de mogelijkheid om de tijdstap te halveren of verdubbelen afhankelijk van het convergentiecriterium van de oplossing, waardoor de tijdstappen aan te passen om de rekentijd te beperken, en de mogelijkheid om te beginnen vanuit een situatie met droge cellen.

2. Het ontwikkelen en testen van een gekoppeld 1D-2D software systeem voor het simuleren van de wisselwerking tussen ondergrondse afwateringssystemen en bovengrondse vrije oppervlaktestromingen.

3. Ontwikkeling van een efficiënte manier voor het simuleren van stedelijke overstromingen op het gewenste grove resolutie rooster met een 2D software model dat gebruikt maakt van informatie verkregen uit beschikbare topografische

data met hoge resolutie, zodanig dat de afwijkingen in het simulatie proces minimaal zijn.

4. Het opnemen van infiltratieprocessen in de software voor 2D vrije oppervlaktestromingen zodat het effect van maatregelen voor het beheersen van de piekafvoer kan worden beoordeeld.

Dit proefschrift eindigt met aanbevelingen en suggesties voor aanvullend onderzoek om de ontwikkelde methode verder te verbeteren en geeft een aantal algemene overwegingen voor het modeleren van overstromingen in stedelijke gebieden. De belangrijkste aanbevelingen voor aanvullend onderzoek zijn:

- Het beter bepalen van de effectieve ruwheidcoëfficiënten die de verandering van de ruwheid binnen het grove rekenrooster nauwkeuriger weergeeft.

- Het beter weergeven van berging en stroomoppervlak binnen het grove rekenrooster in de vorm stroomafhankelijke berging-diepte en stroomoppervlakte-diepte relaties.

T - #0675 - 101024 - C0 - 240/170/9 - PB - 9781138000483 - Gloss Lamination